Beginner
+
Intermediate Guide to Japanese Joinery

Jin Izuhara

© Copyright 2019 - All rights reserved.

It is not legal to reproduce, duplicate, or transmit any part of this document in either electronic means or in printed format. Recording of this publication is strictly prohibited and any storage of this document is not allowed unless with written permission from the publisher except for the use of brief quotations in a book review.

Table of Contents

Book 1: Beginner's Guide to Japanese Joinery

Introduction 6
Chapter One: What Is Japanese Joinery? 14
Chapter Two: Japanese Joinery and Techniques 22
Chapter Three: Japanese Wood And Tools 49
Chapter Four: The Process of Japanese Joinery 60
Chapter Five: Building a Simple Square or Rectangular Frame 74
Chapter Six: Construct a Step Stool 79
Chapter Seven: Create A Tool Chest 118
Chapter Eight: Make A Dynamic And Beautiful Dinner Table 132
Chapter Nine: Lacquering and Preserving a Wooden Object 151
Chapter Ten: Japanese Joinery And Taoism 163
Final Words 167

Book 2: Intermediate Guide to Japanese Joinery

Introduction 169
Chapter One: Japanese Joinery, Tools and Workspace 172
Chapter Two: Advanced Joints 188
Chapter Three: Project 1—Simple Stool 215
Chapter Four: Project 2—Three-Shelved Rack 219
Chapter Five: Project 3—Cupboard with Two Shelves. 224
Chapter Six: Project 4—Dining Table and Chairs 230

Chapter Seven: Project 5—Dynamic Bed with Two Dressers 246
Chapter Eight: Maintenance of Wood and Japanese Lacquering 257
References..266

Beginner's Guide to Japanese Joinery

Introduction

Welcome to **A Beginner's Guide to Japanese Joinery.** Japanese carpentry practices were developed more than a thousand years ago. These ancient methods of joinery for building and furniture-making are complex and exquisite, yet incredibly functional without the need for glue or nails. The art of Japanese joinery has remained in practice through the centuries by highly trained craftsmen using intricate, hand-held tools. The purpose of this book is to introduce you to this fascinating craft and share some of the key skills you can develop to begin your training journey.

Throughout this book, you will learn about the theory, the specialties within this ancient profession, and the materials traditionally used. You will also learn about the different types of joins synonymous with Japanese carpentry and about the tools required to create them.

Within this book are four practical exercises that will lay the foundations for your journey into mastering Japanese joinery. You will begin by practicing a simple task and move forward into basic furniture items to hone your skills. Each exercise will prepare you for the next stage of your training.

To begin, it is vital to take some time to explore Japanese history and culture. These facts will help you understand this country's elegant belief system surrounding natural materials and building practices. Identifying with the country's history will also reveal how these unique techniques came to develop in such a different manner to western woodworking methods and what events led to these advancements in carpentry.

Japan is an island nation located in the northwest Pacific Ocean. The west coast of the country faces the Sea of Japan and extends from the Sea of Okhotsk in the north to the East China Sea and Taiwan in the south. Japan comprises an archipelago of 6,852 islands within the Pacific Ring of Fire. The five main islands from north to south are Hokkaido, Honshu, Shikoku, Kyushu, and Okinawa. The capital city of Tokyo is located on the central Pacific coast of Honshu.

Japan is a highly populated region with 126.2 million inhabitants. As a large proportion of the land is mountainous, the urbanized areas are mostly coastal, vastly populated, and narrow. Tokyo is the most populous metropolitan area in the world, with 34.7 of the country's people residing in the greater city area.

With around 73 percent of the land made up of dense forests and mountains, Japan has a stunning landscape, mostly untouched by human development. The majority of the land is unsuitable for habitation, industry, and so farming remains conserved for nature and wildlife. A large network of national parks has been established to protect the native wildlife of the region.

Culturally, Japan is multifaceted and rich in history and traditions, which has created a unique development of civilization that sets the country apart from the rest of the world in many ways.

During several centuries of a feudal era from 1185, Japan was characterized by a ruling class of warriors, named the Samurai, governing in concordance with the Imperial Court. This was a military-dominated government system defined in history by invasion attempts from the Mongol armies, attempted rebellions, and civil wars between feudal lords.

Within this era, prosperity was achieved in farming, population growth, and commerce. The popularity of Buddhism, introduced from China centuries before, spread from the elite classes to the general populace, encouraged via the embracing of this religion among the samurai.

Within the feudal era trade, it was established with Portuguese traders and Jesuit missionaries introducing European technology and firearms to the nation. The civil war and feuding were of great benefit to the Portuguese who were able to trade firearms with the Japanese armies. Small pockets of Christian colonies emerged with some success in converting Buddhist Japanese locals to the religion. Portuguese became the first western language to receive a Japanese dictionary as the Jesuits hoped this would aid more conversions to Catholicism.

Introducing European weaponry to the warring factions of Japan led to an imbalance of power with rivals seeking to conquer each other. The Japanese were also inspired to launch two failed invasion attempts of Korea. After an open war broke out between rival clans in 1600, the ruling Tokugawa shogunate began designing measures and codes of conduct to control the rival factions and create political unity. This included strict penalties for social unrest, often harsh executions, and outlawing the practice of Christian religions, foreign books, and other western practices. Thus began the Edo period.

In 1639, these measures for peace and unity led to a significant decision that would greatly impact the development of Japanese culture for the next 200 years. This was the year the Tokugawa dynasty decreed Japan an isolationist state. This meant closing the country to further foreign influence that could potentially cause dissent among its people. A single trading post on the island of Dejima was allowed to remain open to the Dutch, who were the only Europeans able to step foot on Japanese soil. The country continued to trade with China and Korea, but the Japanese people were forbidden from building ocean vessels or traveling abroad. Any Japanese person who did were not permitted to return.

The Edo period created an encapsulated society within Japan, shielded from any outside influence. Where many other cultures of the world at this time were within a phase of exploration, cultural influence, and growth, Japan was focused on maintaining and developing its own unique culture.

This isolated period in Japanese history did not diminish cultural growth; instead, the opposite occurred. Numeracy and literacy flourished in both urban and rural areas, as schools were often attached to local shrines. Art and entertainment advanced, and a vast commercial publishing industry thrived. With less fixation on feuding, the merchant classes, growing in wealth, gained interest in social pursuits such as theatre and music. The unique elements of Japanese culture stemmed from tradition and history, were focused on and developed without changes caused by outside influence.

The Edo period also affected architecture for public buildings and dwellings. During the Feudal era, the Shinto tradition of building around gardens was adapted to suit defense. For example, spaces designed for gardens were re-purposed for training practices in the preparation for battle. Untraditional stone and brickwork were incorporated to protect important locations from potential attacks. As unity grew and fighting diminished, the popularity of Buddhist and Shinto influenced buildings once again became popular.

As the populace grew, dwellings with two stories were more common.

Traditional Japanese architecture was heavily influenced by nearby Asian countries, especially China. Although several styles and geographical differences created attributes unique to Japan. The buildings synonymous with the country are connected to the two main religions—Buddhist temples and tea houses, and Shinto shrines.

The most common and vital building material within Japanese architecture is wood. Japan has a vast amount of forests, and this has always been a key resource utilized for building. Wood is a deeply honored material in Japan as the Shinto belief system encompasses profound respect and worship of nature.

Shinto is a belief system that reveres nature. The religion's spiritual practices and rituals are based on the belief that the Kami (spiritual beings connected to elements of nature) are embodiments of the power of nature. Nature is worshipped, and a balance is sought between humanity and the mysterious power of nature to create a mutual harmony. Shrines have been traditionally built as sites of spiritual importance where this worship can take place. Collectively, these shrines are viewed as an interconnected web of entry points to the Kami, where human worship can be communicated.

The reverence of nature inspires an elegant relationship with wood as a building material for the Japanese that is unique to this culture. The forest is sacred, so wood is treated with huge respect by carpenters and craftsmen. Within Japanese joinery, there is terminology for joins made, using different elements of the tree. Yukiatsugi is the joining of two ends taken from the top of the tree's trunk. Wakaretsugu is the joining of two ends taken from the base of the tree's trunk. Okuritsugu is the joining of a piece from each of these parts of the tree.

Japanese architecture has also been greatly shaped by geographical disasters. The Japanese archipelago is positioned within the Pacific ring of fire, a horseshoe-shaped area located within the Pacific Ocean where several tectonic plates meet. The ring of fire is the most active earthquake belt on Earth, making the area also prone to underwater volcanoes and subsequent tsunamis.

Destructive earthquakes, often resulting in tsunamis, tend to occur in Japan several times each century. Whereas in low earthquake risk areas of the world, humans naturally developed building practices using earth and stone, the Japanese found that wooden structures built with the use of complex joins had a better chance of surviving the frequent disasters of their region.

Japanese joinery has been developed to withstand and counteract the level of damage from earthquakes, but also to create incredibly strong building frameworks capable of bearing huge weight. Wood is carefully selected and often aged within these practices. Woods are also selected for the beauty and quality of the grain, as nature is respected and is a very present symbol of Japanese cultural themes. The Cypress Hanoi has been the most sought and used wood for more than 1,000 years.

Traditional Japanese buildings feature a structure of posts and lintels supporting a gently curved roof. The inner walls are never load-bearing, often paper-thin, and generally movable. The internal layout with adjustable screen walls allows for the space to adapt as required for the purpose. These types of walls, named shoji, also mean less destruction from earthquakes and are more easily replaceable. The roofs of these structures are wider than the internal space from all sides. This requires a complex bracket system of beams known as Tokyo. This bracket system is more elaborate in shrines and larger buildings, and more simplified in domestic buildings.

The relationship between the inside and outside is somewhat interchangeable for public buildings and homes as the garden and nature are seen as important, according to Shinto beliefs. The interchangeable moving walls incorporate this relationship for ease of access and views into the garden or surrounding natural environment. For temples, this feature opens the building for the presence of visitors.

Unlike western buildings where the structural components are usually hidden beneath plaster or paneling, the structural elements of Japanese buildings are displayed. These wooden frames, beams, and posts are a sublime blend of form and function that add to the overall beauty and ornamentation of the building.

Both internally and externally, Japanese buildings aesthetics adhere to a minimal and simple but elegant style. These visual ideals stem from Shinto and Chinese Taoism. Natural materials are present, not only in the wooden structural elements, but also in the rice straw mats, paper and silk wall treatments, and bamboo screen frames. This idea of minimal and natural beauty is still present today in contemporary Japanese architecture and interior design.

Japanese joinery techniques have also been influential throughout the country's history for furniture design. The themes of elegance, functionality, simplicity, and minimalism are also transferred onto the interior spaces. Negative space is viewed as being important as the necessity for functional lifestyle items, so every item is deliberate and exact. As furniture is traditionally crafted from wood, the intricate joinery for strength and function are as vital as for structures. The interchangeable nature of interior spaces adds an essential function to furniture—the ability to be easily moved between spaces. This lifestyle factor has created several design features to be common within the design of these items, such as handles for lifting, minimal decorative effects, and lightness for movement.

Chapter One:
What Is Japanese Joinery?

Japanese carpentry is renowned globally for its highly refined craft and precision. The delicate balance between complex technique and elegant simplicity can best be described as 'geometry meets nature'. In this chapter, Japanese joinery will be explored in detail, delving further into the history, the techniques, and some examples of traditional Japanese structures synonymous with this building style.

Having an interest in woodworking connects a person to a long history of craftsmanship spanning human civilization. Many people begin this journey during their school years, learning the basic skills required to complete simple tasks, become comfortable with physical tools, and invoke interest in this activity. This study may lead to a lifelong enthusiasm for carpentry as a hobby, or potentially a profession.

Wood is a natural material, and its response to different tools and crafting methods can be understood from considering the tree. Trees grow upward and outward with each season marked with rings that are visible within the grain as sections are cut. The growing pattern of the tree needs to be considered in terms of directionality for cutting to get the best performance from the wood. The Japanese profoundly respect the tree, as well as the wood. To follow Shinto practices, the tree informs building decisions. For example, wood taken from the south-facing part of the tree's trunk will be reverentially used in the creation of the south-facing side of the building or shrine.

Understanding the relationship between wood and the craftsman is of huge value to anyone working with this material, like any other physical material. Cultivating this relationship through practice and experimentation will inform dexterity, ability, and accomplishment. No piece of wood is the same; the grain is as unique as a human fingerprint. The Japanese select each piece of wood thoughtfully and deliberately, and this is a beneficial lesson to apply to any kind of woodworking practice.

Many people incorrectly believe that woodworking can be hard labor and difficult when using hand tools. However, when tools are properly sharpened and cared for, the tools will do much of the hard work without a lot of force at all. Sharpening chisels, saws, and planes are especially important for Japanese joinery where precision is vital, and too much force could result in destroying a day's work on a complex range of shapes.

Before attempting Japanese joinery, it is advisable to already be familiar with the tools and methods used for cutting and shaping wooden objects. Coarse tools are best for coarse work, and fine tools are best for fine work.

It is also important to have a suitable space and basic workshop equipment available. This need not be any more than a simple workbench in a garage or shed. Having a designated space will help to create a suitable headspace for concentrating on the tasks to be achieved. When focusing on fine work, as is needed for joinery, comfort and free movement are key. Therefore, it is useful to test where the best place to stand will be, whether a stool is required, and where wooden pieces can be clamped in place for precise cutting. It is a Japanese custom to sit on rice straw mats on the floor, but this may not be comfortable for everyone.

In modern architecture, materials are factory cut and fabricated to exert as much control as possible to achieve precision and replication. Ancient Japanese craftsmen, the Shokunin, did not need such machines to achieve the same precisely repeated shapes and elements. Many of the world's oldest surviving wooden structures can be found in Japan, created with joinery techniques.

The shokunin did not select wood for uniformity; instead, they selected it by using the unique attributes of the tree as an advantage. Inconsistencies were masterfully balanced and counteracted, and the entire life of the wood was utilized. Shokunin are able to understand and anticipate the behavior of the wood over time—how it would expand and contract depending on climate variation, how it would inevitably shrink with age, and which position would suit it best due to its previous living state as the tree.

These skills were often the difference between a building surviving an earthquake or being destroyed. The results show a sophisticated technological mastery that is astounding for the period of history in which they originated. These methods of manufacture are far more robust than many modern ones in use today. Modern manufacture also does not prepare for the same level of longevity, ease of replacing any damaged components, and potential for disassembly with minimal waste or environmental corruption.

The fundamental aspect of Japanese joinery is that only one material is used or required. The shokunin understood that wood has both the strength and flexibility to complement how it is joined, and the join will be long-lasting. In Japanese joinery, the strength of the wood is used to both add and counteract its own weight and pressure in an elegant balance that is stronger than any nails or adhesive could be.

This is done by cutting complex geometric shapes into the ends of the wooden pieces that must be joined. These shapes are countered and symmetrical with each other, the negative space of one mirroring the physical space of the other precisely. The pieces are designed to neatly slot together, sometimes using an additional pin-shaped piece slotting from an alternate angle. The direction of joining is the opposite of the pressure of weight that will be placed on the join, meaning that once interlocked, the two pieces are immovable.

These methods adhere to the Buddhist and Shinto principles of respecting nature. As the tree must be cut down to build the structure, the carpenter owes a debt for the life of the tree. By using the wood to create something both long-lasting and beautiful to behold, that debt is settled. It is intended that the structure will exist for as long as the tree may have lived untampered with. By fulfilling this ideal, the spirits connected with trees will be appeased.

An important example that demonstrates the astounding skill of the shokunin, and the spiritual principles applied to the building, is the famous Horyu-Ji Temple. This temple, located in the ancient capital of Nara, is the oldest wooden structure in the world. The temple was originally built in 600AD, then re-built in 700AD after a lightning strike caused a devastating fire. For context, this was the time frame in which the Mayan empire was prospering, the Anglo Saxons were reclaiming the British Isles from the Roman Empire, and the Chinese were discovering the formula for gunpowder. At this point, the Japanese were designing and building structures with wood that would last more than a thousand years and still be going strong today.

The Horyu-Ji, a temple in Irakuga, Nara Prefecture, Japan. Image by RPBaiao

The Pagoda is the oldest building within the temple complex and now also serves as a Buddhist museum, housing many of the religion's most priceless Japanese objects and educating visitors on the story of Buddhism in Japan.

It was built using 2000-year-old Japanese cypress (hinoki). The trees were cut and prepared using handheld tools, nothing like the saws and machinery used today in harvesting lumber. The carpenters also carefully selected the wood, understanding how each part would behave due to the tree's position, angle toward the sun, and quality of the soil where it was located. All of these factors were considered and taken into account for the wood's likelihood to expand or bend with moisture, which direction a bend may occur, and any potential for rot. Each piece of wood was then designated a suitable placement within this complex structure that would make the most from its unique characteristics and would demonstrate respect for the tree that was sacrificed.

The temple complex is reflective of a traditional Chinese Buddhist monastery plan. It is laid out on a north-to-south axis, with a south-facing entrance. In addition to the Pagoda, the complex includes the main hall, lecture hall, north gate, and the Great South Gate. The complex is surrounded by a walled corridor with a colonnaded interior and walled exterior. This cloistered walkway has several Chinese features, such as wood columns, window openings, and plaster exterior walls.

The complex differs slightly from traditional Chinese design in that the Pagoda and Kondo (lecture hall) are offset, instead of symmetrical placement. This is believed to be for a visual purpose so that all of the structures can be viewed simultaneously from various angles when approached by visitors.

Temples and shrines in Japan used to be the tallest structures in the country. The Horyu-Ji Pagoda is five stories tall and stands at 122 feet in total. The structure is partially supported by one central column rising the entire height. Outer columns with cantilever brackets support the five roofs. Each roof surrounds the structure, curving upward and inward, and these roofs diminish in size with each floor to enhance the height of the structure. To support these roofs and their heavy tiles, complex framing brackets with interlocking joins were constructed. Joinery techniques were also utilized for the staircases, floor platforms, inner walls, railings, banisters, window frames, and the plethora of ornamental wooden details within the building's interior and exterior.

The design of the landmark curved roofs of the temple and other traditional buildings is a definitive Buddhist style. The curvature of this roof shape is a complex structure to realize without mathematical input. The slope and turned up eaves require a geometrical system for ensuring repetition and symmetry. Without knowledge of the mathematical equations required for exact repetition (which were not introduced until the 18th century), carpenters worked by eye and with practical measurement methods. Line drafting and a standard carpenter's square were used to drawing the angles and lines for cutting on the surface of the wood.

The posts' and beams' structural style used for Buddhist temples and other traditional Japanese buildings is known as a 'rigid frame' structure. The type of joinery predominantly used for securing the posts and beams are mortise and tenon connections, with the use of wooden pegs and wedges that slot into the two pieces. This method, whilst very strong, provides a level of flexibility within the structure that allows for expansion and contraction of the wood due to climate and also for preventing damage from earthquakes.

This famous landmark gives an idea of the incredible skill that Japanese carpentry requires, and its uses in practice for the joinery that this book describes.

In a nutshell, Japanese joinery is all about how to make joints with hand tools such as a chisel, hammer, mallet, saw, and others without using any fastener like screws, nails, and even glue.

Chapter Summary

In the next chapter, the techniques of Japanese joinery will be explored further, illuminating the different styles, how they are created, and for what they are typically used. However, this chapter dealt with the meaning of Japanese joinery from the pre-history time to present-day Japan and how it has affected wood construction in its entirety.

Chapter Two:
Japanese Joinery and Techniques

In this section of this chapter, the different styles of Japanese joinery will be explored in greater detail and explained with examples of what type of structure the join is traditionally used to build. A suitable way to envision and describe the intricate joints of Japanese joinery is by considering it as a three-dimensional puzzle that is simple from the outside, yet complex from within. Once the mechanism is understood, the puzzle is easy to solve.

The joints described in this section can be created with a range of complexity. For a beginner, the joints can be constructed in a manageable and uncomplicated manner. Once the basic skills and understanding have been mastered, more complicated joints can be explored.

Before investigating these joins, it will be helpful to become familiar with the terminology used to describe them. The guide below is a helpful reference to learn and follow these words within the descriptions:

- **Tenon:** This is a protruding piece of wood designed to fit into a slot (mortise).

- **Miter:** This is a cut made at 90 degrees into the main surface intended for a join bisecting this angle to form a corner.

- **Rails:** This describes horizontal lengths of wood in frame and panel construction.

- **Stiles:** This describes vertical lengths of wood in frame and panel construction.

- **Spline:** This is a strip of wood that is inserted into grooves cut in the edges of the interior corner of a frame.

- **Dado:** This is a slot cut into the surface of the wood to hold and counterbalance a shelf.

- **Mortise:** This is a hole or recess cut into the wood intended for a corresponding part (tenon) to slot into.

- **Bevel:** This is a tool used to layout and transfer angled lines for a miter cut.

A simple and effective method of making joints without using nails or screws

Here, we consider the mortise and tenon joint, which seems to be the oldest ways of joining two pieces of wood. It consists of a mortise hole and tenon tongue that fit into one another. As one of the versatile joints, mortise and tenon can be applied in many different woodworking types including fine furniture, framed buildings, and many others.

To make this particular joint, the following tools are required, including tenon saw, mortise chisel, steel rule, mortise gauge, marking knife, steel square, pencil, and clamp. Just follow these steps.

Step one: Prepare materials for the two parts of the joint.

Step two: Mark the sizes on the wood. Use 'Rule of Thirds' to set out the sizes. What this rule implies is that the proportions of the joint will be divided into three parts. So, if the tenon rail measures 20mm in thickness, the mortise chisel should be almost one-third of it, which is approximately 7mm.

Step three: Adjust the mortise gauge pins to the width of the chisel.

Step four: Make two arrowed dots and examine each side to ensure the tenon is centrally placed.

Step five: Mark the depth of the tenon to be almost two-thirds of the depth of the second piece of wood. It should form the shoulder line.

Step six: Mark parallel lines all around while holding the wood in the vice.

Step seven: Set marking gauge to 6mm and measure line all the way around.

Step eight: Set a marking gauge to the wood's width, which is 6mm, and mark the width of the tenon.

Here comes the completed marking out of tenon end grain.

Step 10: Lay tenon wood piece on mortise material and mark the position of mortise using a pencil. Then, transfer lines to the wood.

Step 11: Set mortise gauge to mark the position of the mortise, and do not change the distance between the pins. You can mark the width of the mortise with the knife.

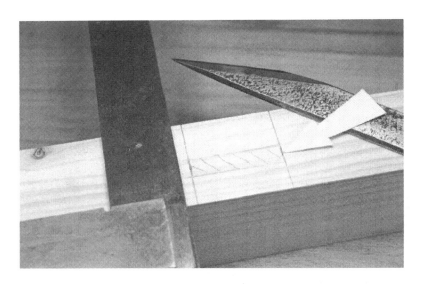

Step 12: Clamp the mortise position over the bench and chop off the mortise. Stand at the end of the bench and align the chisel vertically when chopping the mortise.

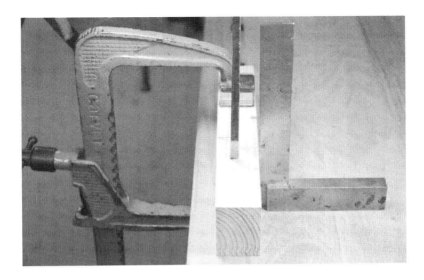

Step 13: Chop off the waste with the chisel as a lever to remove the chippings. Stop at 2mm from the knife line.

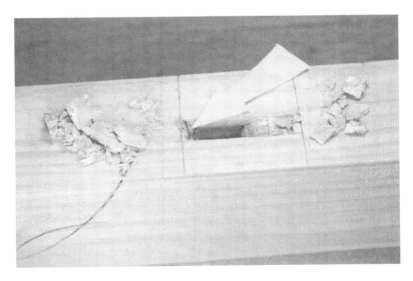

Step 14: Check to see if the depth is 3mm longer than the tenon.

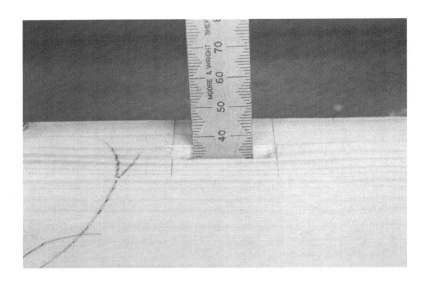

Step 15: Make two final cuts, one at each end on the knife line. Check it with the square.

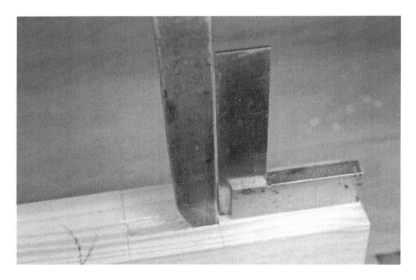

Step 16: Cut the tenon by hand using a hand saw to make two triangular cuts, one for each side.

Step 17: Finish the cutting by sawing in the horizontal direction to the shoulder line.

Step 18: Make two additional cuts on the face of the board.

Step 19: Position tenon piece on the bench hook and saw off cheeks from tenon all round, leaving 0.5mm waste from the shoulder line.

Step 20: Place the wide chisel on the knife line, look for the square, and chop off the shoulder line. Then, use the block plane to chamfer the end of the tenon.

Remember, the joint should push together with moderate hand pressure applied as seen in the picture below.

Special characteristics of various joints and their specialized use

Presented below are the various joints with their unique characteristics, as well as specialized uses.

Interlocking Tenon Joint

The interlocking tenon joint is generally used for the creation of staircases and making chairs. Three pieces of wood can be joined together in a cross shape, such as the seat platform joined to each leg of a chair. The leg features a cross-type cut, leaving space for two rails to the slot and lock together placed from above. The rails feature a central square-shaped cut designed to affix the two pieces within the interior of the cross-shaped cut-out of the leg.

Interlocking Miter Joint

The interlocking miter joint is popular in heavy frame construction. Two pieces of wood are joined to create a right angle. Each end of each piece is chiseled to form a half-lap with mitered shoulders protruding at a 45-degree angle. The two tenons are shaped similarly to arrowheads that overlap each other inside the mortise with a step-type shape. The two pieces slot together leaving a slot open for a thin spline that secures the join in place and prevents movement.

Three-way Corner Miter Joint

The three-way corner miter joint is often used for tables, such as desks and dining tables. All three pieces feature miters, and the table leg has a tenon that fits into notches cut into the other two pieces (rails). The leg's end forms a pointed triangular shape on the outer side fitting into a corresponding shaped mortise on each rail. On the inside, a square-shaped tenon cleanly covers the join within. For thicker rails that require a sturdier reinforced joint, a three-way pinned corner joint is very similar, but the leg section features two tenons acting as pins that slot into the two rails.

Full Blind Dovetail Joint

The full blind dovetail joint is also sometimes known as the secret mitered dovetail joint. These types of joints are used in furniture-making to join the sides of a cabinet or box. Of all types of dovetail joints, these are considered the strongest. Alternating angled grooves, similar to teeth, are chiseled from the interior sides of each piece of wood to be joined. A lip is left on the exterior walls of the wooden panels that will meet as the dovetails join, disguising the complex teeth. The two sides then neatly slot together in a joint that is invisible from the outside.

Sliding Dovetail Joint

The sliding dovetail joint is often used to join rails to legs in chair construction. This joining technique features tenon and miter shapes that slide together with a stabilizing tenon to reinforce the join. The legs and rails come together forming 90-degree angles, suitable for a seating platform.

Shelf Support Joint

The shelf support joints are designed and used for shelves that can bear heavy loads. A dado is required to be cut into a vertical wooden section to support the horizontal shelf. The dado has blind tenon cut-outs corresponding to the shapes cut into the horizontal shelf to match the stopped dado. The tenons can be of varying shapes and sizes, depending on what is best for the shelf length and type of panel it will be joined to. The join must be tight to support the weight of the shelf and load.

Divided Mortise and Tenon Joint

The divided mortise and tenon joint is used for large frame and panel pieces used in the construction of doors and other features within buildings and furniture, such as cabinets. Tenons are cut into the rails to fit through the stile's mortise. The rails often join together through the stile, combining three pieces in a cross-type shape. When this is the case, the tenon and mortise join the two rails in the slot together within the opening of the stile, neatly hiding the join.

Mitered Shoulder Tenon Joint

The mitered shoulder tenon joint has a similar purpose as the divided mortise and tenon joints for the construction of frame and panel structures and objects. The difference is that the surfaces bordering the mortise and tenons on one side are beveled, creating angled joining shapes. This creates a pleasing "X" shape where the three wooden pieces meet, which enhances the unique attributes of each piece of wood and their grain patterns.

Mitered Corner Joint

The mitered corner joint joins two pieces of wood together to create a 90-degree angle. The join itself forms a 45-degree angle within the corner's interior. A concealed dovetail tenon in a triangular shape is cut from one edge to slot into a matching mortise on the other edge. The two pieces are combined with no visible evidence of the internal join. This type of join is typically used to create a large frame, such as a sliding door frame, or window frame. This join can be created with one tenon, or more than one by repeating the same shape, or asymmetrically cutting the triangular shapes and their corresponding mortises into the two wooden pieces.

The above Japanese joins are the basic methods used in ancient Japan, as well as today by carpenters all over the world. Using these joining designs, any manner of woodworking projects can be undertaken, from furniture to the structures of buildings.

Although it may have seemed unfathomable that the shokunin were able to achieve incredible large scale projects such as the Horyu-ji Temple complex using simple hand-made tools, breaking down the individual techniques they used aids in understanding. Every carpentry project is simply a series of small tasks performed one by one until the vision is realized and complete. Later in this book, these techniques will be put into practice via a series of tasks following small steps to accomplish a finished object.

Three joints that don't need any specialized tools, and how to make them

Here are three main joints that you can use to make virtually everything from cool shelves to rocking chairs.

Sliding Dovetail: The sliding dovetail is used for braces on doors. It is equally used to attach chair legs to a chair or stool seats, as well as chair legs to chair feet.

How to make a sliding dovetail

The first step is to lay out the dovetail so that it becomes narrow in both directions. The suitable edge angles range from 3 to 5 degrees.

Secondly, cut the dovetail out, sand or plane the edges until they become straight and smooth.

Thirdly, use the dovetail sides to protect the saw at the right angle in order to cut the dovetail groove.

The fourth step is to chisel out the groove and set the dovetail until it gradually slides in and fits well.

Lastly, you now have a completed sliding dovetail. The sliding dovetail can be used to create a leg joint on a work chair that becomes tight as more loads are released on it.

Dado Groove: A groove joint is always used to take hold of the ends of shelf boards to make them stronger. Also, it can be used to catch chair seats on legs, as well as catch splines between boards. It is useful anywhere you want to keep a board from twisting or bending or even make a stronger joint that will carry more load.

How to make a dado groove

Layout the groove, marking it 0.125-0.5-inch deep, based on how much load it can carry.

Now, clamp a square-cut block to the board to keep your cut in a perpendicular direction.

Saw both sides, hammer, and chisel down to the bottom line.

Complete the dado groove. However, a very simple way to cut the dado groove is to use a table saw with a dado blade.

Dowel Joint: Dowels come in various sizes and can be used to join boards side to side to create workbench countertops, chair or stool seats, and larger shelves. Apart from that, dowels can also be used at the end of a board to incorporate them into an upright for shelves.

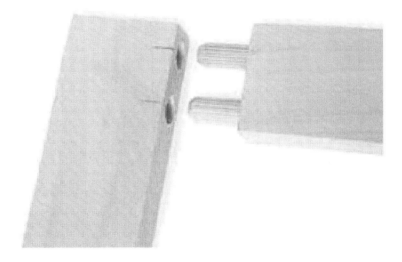

How to make a dowel

Half-inch to three-quarter-inch dowels connect the boards together and are held together by 0.25-inch lock dowels.

Since dowels are usually dry and wood tends to be somehow wet, they usually swell up and stay in place without glue. So, you may call it a permanent joint that will not break easily. However, you will have to drill out the lock dowels to break or separate it apart.

Chapter Summary

Every woodworking project requires the correct wood, and tools to handcraft the best possible finished item. In the next chapter, the wood that Japanese craftsmen select and prefer will be explored. The traditional tools used by the shokunin will also be described with examples of how they are used in practice creating the joinery explained in this chapter.

Chapter Three: Japanese Wood And Tools

The Japanese's respect and admiration for nature is a core theme within Japanese carpentry, in culture and methods. The wood chosen for building and furniture-making is selected for its specific qualities, appearance and symbolism. In this chapter, the types of wood predominantly used in Japanese carpentry will be explored.

As the islands within the Japanese archipelago feature an abundance of forests, wood was the dominant material used in ancient building methods. In this area of the world, there are native tree species, and the Shokunin discovered which of these species were most suited to different types of projects.

The skills of these craftsmen also encompassed a deep understanding of how each of these species of woods would behave when transformed into lumber. Over centuries of practice, the shokunin cultivated knowledge of how many factors would affect the wood, such as climate, location and position of the tree within its environment.

There are several species of tree that the Shokunin favored and are still used heavily today. These are Japanese Cedar (Sugi), Japanese Red Pine (Akamatsu), and—the most popular and respected—the Japanese Cypress (Hinoki). These different woods have traditionally been used for specific types of objects in buildings or furniture.

Japanese Cedar

The Japanese Cedar is the national tree of Japan and is often planted around temples and shrines. This is a very large evergreen tree that can reach 230 feet in height and 13 feet in trunk diameter. The bark is a deep, reddish-brown color, and peels in vertical strips. The leaves are spiraled and needle-shaped. The Cedar prefers a forest in a warm, moist climate with well-drained soil.

The Japanese Cedar produces a fragrant timber that is soft with a low density. In building, this wood is found to be particularly weather repellent and resistant to insects and decay. The color is a warm pink with an attractive grain pattern that ages well.

This tree is popular in many types of Japanese construction due to its versatility. The strong, yet light, wood is easy to cut. For these reasons, Cedar is used for furniture-making and other indoor projects, such as paneling and pillars. If buried for aging, the wood turns a beautiful green that is coveted and increases its value.

Japanese Red Pine

The Japanese Red Pine grows throughout Asia and the southeast of Russia. This tree is popular for both timber and as an important ornamental feature for gardens in Japan. The Red Pine is a slender, tall tree that can reach up to 114 feet in height. The needle-shaped leaves are green but can turn yellowish seasonally. The tree prefers well-drained, slightly acidic soil in full sunlight.

The Red Pine's heartwood is a light reddish-brown in color, and the sapwood is a much paler yellow. The wood's grain is straight and linear with an even texture. With a high resin content so it is slightly oily to touch, this wood is lightweight, strong and resistant to rotting.

The Red Pine was traditionally used in bridge construction, due to its unlikelihood to rot over time in wet conditions. It was also popular in temple building, especially for the roof beams.

Japanese Cypress

The Japanese Cypress tree is native to central Japan. It is cultivated by the Japanese for its high-quality wood for timber and ornamental properties. It is worthy to mention that many variations have been bred for differences in size, branch spread and leaf style. It is a slow-growing tree that can reach 115 feet in height and around 3 feet in trunk diameter. The leaves are long and green with blunt tips.

The timber from Japanese Cypress has a pleasant lemon scent and is a light pinkish brown. The grain is straight and beautifully rich. The Cypress dries quickly, which reduces warping, is durable over vast amounts of time and is resistant to rotting.

The Cypress was traditionally used in many building capacities, importantly the construction of temples, shrines and palaces. This versatile timber was also utilized for furniture-making, notably for baths due to its fragrant scent. The Cypress was vital within religious practices, used for Shinto ceremonies and purification rituals.

Japanese Cypress is not only valued as a great building material, but it is also mentioned in ancient Shinto texts as the 'sacred tree'. The wood from the Cypress has many benefits aside from its strength and durability.

All trees emit a substance called phytoncide, a heady aroma with the function of repelling insects and bacteria. The phytoncide scent released by the Japanese Cypress is incredibly pleasant and also causes positive physiological responses in humans. Exposure to this scent can lower blood pressure, improve immunity, have a calming effect, and aid the heart rate and nervous system. The extracted oil from the Cypress has antibacterial properties utilized for the treatment of MRSA (Methicillin-resistant Staphylococcus aureus).

Cypress wood contains aromatic compounds called terpenes, known to absorb toxic substances, such as formaldehyde. The use of Japanese Cypress for buildings has the added benefit of controlling ticks, preventing insect infestation and molds. The Cypress has enormous benefits within temples and homes creating a harmonious and safe environment for residents and visitors. For this reason, it has become common practice for some modern Japanese construction companies to infuse tatami (rice straw) floors, flooring plywood and wallpaper with Cypress oil.

Due to the continuous demand for the Japanese Cypress over the centuries, and the length of time required for the trees to reach a harvestable size for building materials, the production of logging forests have been pushed to the limits. It is part of Shinto belief practices that the many shrines all over the country must be removed and rebuilt every 20 years to please the spirits they are attributed to. This practice symbolizes religious renewal and is a vital part of traditional culture. The rebuilding of these shrines requires 10,000 Cypress trees each 20-year cycle, but the trees take 400 years to grow to the necessary height and size to accomplish these projects.

Mass deforestation in Japan and environmental damage have added to the increasing loss of viable Cypress. The International Union for the Conservation of Nature has listed the Japanese Cypress as 'near threatened'. This could raise the demand for the wood, as the limited availability will define it as a luxury material. The Kiso Valley is home to the last sizable Hinoki forests left in Japan. These forests are managed by the forestry agency, which has begun conservation efforts to manage the number of large trees allowed to be felled each year. These measures are also intended to protect the forest's ecosystem.

Japanese Carpentry Tools:

The distinct woodworking joints of Japanese joinery are created with handheld tools that can be traced back to China in origin. Japanese tools are crafted from very hard steel that are high in carbon content. The edges for cutting are sharpened to an exceptional keenness that can deliver an incredible smoothness in cuts and plane work.

The immense sharpness of the bladed tools means that they are delicate and require practice. Too much force or incorrect use can lead to the breaking of the fine points and edges. The success of any tool in practice lies within the physical relationship of the human and the tool. Habits of movement need to be developed and practiced before the best results can be achieved. Traditionally, Japanese carpenters practice and train for years as apprentices to hone their skills before earning full responsibilities in their profession.

The principles of Japanese carpentry tools are the same as western tools. There are axes, and saws, and chisels for cutting, planes for refining, boring tools for creating holes, and so on. However, the shapes, sizes, materials and methods vary, as these tools have been developed without western influence for the specific techniques unique to Japanese construction.

In this section, Japanese joinery tools will be described and their uses explained.

Japanese Saw (Nokogiri)

The Japanese saw differs from standard western sawing methods in that the Japanese cut on the saw's pull stroke rather than the push stroke. This means that the blade is designed thinner than western saws. These saws have two main kinds of cutting teeth, the crosscut style (yokobiki), and rip style (tatebiki). These two types of teeth are used in single-edged saws and combined for one type of saw called the ryoba (duel edge). There are different types of saws for performing different tasks.

- **Douzuki:** When saws are made with a stiffening back piece, they are suitable for cutting finer joinery designs; these are called douzuki (attached trunk).

- **Osae-biki:** Saws used for flush cutting pegs without marring the wood's surface are called osae-biki (press cut saw), the teeth of this saw have no set to one or both sides.

- **Azebiki:** Saws for cutting in confined areas are azebiki (ridge saw), this saw is short and round, and has both crosscut and rip teeth.

Japanese Plane (Kanna)

The Japanese plane usually resembles a wooden block (dai) featuring a laminated steel blade, a sub-blade, and a securing pin. The main blade is fixed into position with abutments cut into the sides of the dai. This is a similar method to western planes, although these usually feature a wooden wedge that can be tapped down to position the blade. In Japanese planes, the support bed for the blade is convex rather than a flat surface. The blade is tapered in width and thickness. It can be adjusted from the side to achieve a uniform shaving thickness.

Like Japanese saws, planes are also operated using a pulling motion as opposed to the western pushing method. Carpenters often work in a seated position, using their body weight for more force.

Japanese Chisel (Nomi)

Japanese chisels are made from laminated steel of varying strength depending on the intended use. The angle for beveling also varies depending on the type of chisel, with angles ranging from 20 to 35 degrees. Mortising and heavy chisels utilize a steeply angled blade, whereas paring chisels have shallower blades. Many of the woods used in Japanese carpentry are soft, therefore the chisels are made with this in mind, and less force is required to achieve the desired results.

There are many different types of chisel used in Japanese carpentry. These include striking chisels, heavy timber chisels and slicks. There is a range of other specialist chisels used for exact tasks in construction and furniture-making.

Japanese Gimlet (Kiri)

A Japanese gimlet is a tool used for boring holes via a rotating blade tip. There are three types of gimlet differentiated by the structure of the tip and how it rotates.

- **Momigiri:** This gimlet rotates in a backward and forward motion. It is held between the palms and rotated, alternating between left and right. The handle is traditionally made from Japanese Hinoki or Japanese white pine and is long and tapered. There are various variations of momigiri with different amounts of prongs forming the cutting tip.

- **Bourutogiri**: This gimlet works on the same principle as a screw. The cutting blade is diagonal and wood shavings are ejected as the tip is turned. The handle forms a T-shape that is gripped with both hands and turned.

- **Kurikogiri:** This is also known as a brace. It consists of a sharp head, attached to a U-shaped iron rod. Different types of cutting heads can be fitted into the chuck via a regulating screw. The round part of the handle remains steady, whilst the middle of the handle is rotated. The kurikoguri is suitable for boring larger holes.

Japanese Hammer (Genno or Gennoh):

There are several purposes for Japanese hammers. One is used in conjunction with the chisel for cutting refined shapes. Some hammers are used for positioning hand blades, some for removing nails, and some for tapping out laminated hardened steel from the base of chisels and plane blades.

Marking and Measuring Tools

There are various tools used in Japanese carpentry and joinery for measuring and marking wood. An important part of any woodworking project is accuracy, and these tools are necessary for ensuring that the placing and size of each cut is correct and that precision is achieved.

The inkpot is used to mark long straight lines. The process involves a thread tied to a rounded piece of wood that has a needle attached at its end. The other end of this thread is passed through the inkpot via a small opening and through the depression containing the ink. Then, the thread is wound around a spool. The ink is stored soaked in silk wadding. The thread is held in the left hand whilst the needle is fixed onto the required place on the wood's surface.

The inkpot is then moved away from the needle until enough thread is unreeled for the correct length of the line. The thread is pulled until tight, then released, marking the wood. Other tools used for marking include the kiridashi (marking knife), sumisashi (bamboo pen), kebiki and kinshiro (traditional single and multi-blade marking gauges).

For measuring, traditional carpenters did not have the same methods as modern tapes and rules. They used a carpenter's square. This is a framing square marked with several units of measurement on each side. The square can be used for measuring and marking lines and marks for cutting that will be accurate with each repetition of a shape.

In Japanese joinery, accurate measuring and marking are crucial as each joining piece of wood must be exact for the join to be strong. Each tool plays an important part in creating the ideal shapes, and each tool must be in great condition to achieve a perfect and accurate join.

For cutting tools' blades, a similar process is used for forging samurai swords. Japanese steel is highly refined and strong. A very hard blade metal is welded to a softer piece of metal in a forge. The softer base metal is intended to absorb shock and prevent the harder, more brittle metal from breaking. This technique creates a harder chisel than western models. This also means a finer edge that is difficult to source outside of Japan.

Blades of Japanese planes and chisels have a unique distinguishable feature, a hollow in their flat side called the ura. The purpose of this hollow ensures a high degree of flatness during sharpening, as when the flat side is polished, it is making contact only with the stone on either side of its width. This enhances the precision of cuts made with the chisel and ensures the plane has smooth contact with the wedge and even support. This hollow also reduces the amount of metal needed and reduces friction as the chisel is driven into the wood.

The sharpening process of these blades by Japanese carpenters typically involves three or more whetstones. The carpenter progresses from the roughest stone and ending at the finest. When practicing Japanese joinery, it is vital to maintain the sharpness of the bladed tools and replace the blades of saws to ensure that the cuts made are accurate, and the most is being made of the elegant tools used for these processes.

Chapter Summary

Really, chapter two was the premise of this book. It discussed everything you need to know about joints as stated below:

- The different types of joints.

- A simple and effective way of making joints without using nails and screws.

- Special characteristics of different joints and their specialized uses.

- Joints that don't need any specialized tools and how to make them, and many more.

In the following chapter, we'll learn the process of Japanese joinery in its entirety. Come along as we embark on this tour.

Chapter Four:
The Process of Japanese Joinery

The process of Japanese joinery is based on a learning system that involves many years of commitment and determination of any individual interested in the craft. The secrets of the Japanese joinery and woodworking craft were passed down from Master to apprentice through an oral tradition. These secrets are the same as the technical skills of woodworking. They varied greatly from one carpentry school to another.

It was not until the Edo period that some of the hidden technical skills of the Japanese joinery craft began to be recorded in writing by government officials as a means to establish a standard for residential construction. Still, the knowledge of Japanese joinery was not recorded much even with the intervention of the government.

Unlike many other traditional joinery methods, Japanese joinery has recently remained a secretive craft amongst closely knitted carpentry families in Japan. The complicated joints are made with exactness and skill, deploying various end, corner and intermediate joints to meticulously cancel the effects of loads and torsions.

Components fit together like puzzle pieces to produce intelligent structures. Of course, these combinations are known to be amongst some of the longest-surviving structures even still today. Japanese joinery does not depend on permanent fixtures, such as screws, nails and glues. Rather, joints are firmly held together using interlocking connections and depend on material properties to withstand forces and pressure.

How the Japanese carpenter uses Japanese joinery

Japanese carpentry is popular for its ability to create everything from temples to houses to tea houses to furniture without the use of any nails, screws or power tools. This is done through a process called joinery.

Also, Japanese carpentry has a long history from many centuries ago. Construction in the western world tends to separate an architect/designer from the constructor. However, it's not so in Japan, as the carpenter is also the architect.

In the same vein, Japanese carpentry looks simple and purposeful. The finished pieces' beauty matches development with nature. Nevertheless, the techniques beneath the structures are different and intricate, as they use joinery ingenuity to construct buildings that do not depend on nails or bolts.

Japanese Joinery

Joinery is a construction process that involves the creation of interlocking joints. In a nutshell, these joints carefully join selected pieces of wood together firmly. Traditional Japanese craftsmanship is evident in many of the different traditional inns or small hotels located in Japan.

Japanese wood joints form the basis of the country's great temples, houses and cabinetry. They're designed with joinery techniques that are still being studied by contemporary architects today. These designs are able to use joinery that does not require steel nails, but still forms sturdy internal structures to big buildings while presenting elegant visual constructions within the room's ceilings.

Traditional Japanese aesthetics included wood as the major building material for a number of reasons. The nation was gifted with a myriad of timber resources and the light weight of the material made it a favorable alternative to stone or brick, as earthquakes regularly destroyed the country's coast and rural areas. With the threat of earthquakes in mind, the joints had also been designed to be able to withstand the jolts with the flexibility that other rigid construction materials could not offer.

These joints matched the Japanese minimal aesthetic that had been inspired by early Taoism, but this simplicity was a deceiving perception. The complicated nature of Japanese joinery has been described most times as *geometry meets nature*. It is indeed an *exceptionally well-thought-out process* of incorporating wood pieces to form stable structures. Japanese architecture was never too far from nature, and the character it added to traditional buildings is an obvious contrast to the westernized monoliths that stand tall in the nation's urban areas today.

Many Japanese carpenters have remained loyal to the old crafts, continuing to use traditional joinery techniques in contemporary furniture and woodwork projects. The country manages to produce supreme joinery and woodwork techniques as well as woodwork tools by honoring the craft of past centuries. This same honor also goes to Japan's lauded architecture design.

Fortunately, the processes and secrets are now easier to understand, meaning the advanced techniques have become accessible for even non-professionals attempting DIY projects.

Joinery in small-scale projects

Although dovetailing is one of the techniques in Japanese joinery, it is effective at joining pieces of furniture together. Also, it is relatively a new technique that is more suitable for small-scale projects. It is traditionally right in Japan for carpenters to perform the dual roles of being both architects and home-builders. But, if Japanese carpenters want to handle big projects, what do you think they can do?

Joinery in large-scale projects

There is a need for master Japanese carpenters to utilize advanced techniques for large-scale projects, such as tea houses, shrines and homes. In essence, these advanced techniques make way for ample load-bearing weight. They also allow for a major construction project without the use of common western materials, such as screws or nails.

However, there's one stark disadvantage here. Japanese carpentry is a difficult skill to learn and master. Besides that, it also needs a larger investment in time. So, it is quite hard to join these pieces together using Japanese joinery. Nevertheless, it is a testament to the mastery that these carpenters possess together with the long Japanese history of many years in intricate woodworking.

The Four Types of Japanese Carpenters

Japanese carpenters adopt the same mode of operation. However, they can be divided into four different types of professions. There are three major factors that control the way these carpenters are separated into different types of carpentry, namely:

- Their level of experience with various forms of joinery.

- The wide variety of joints they create.

- The tools they apply in the creation of joints.

The four distinct similar but different professions are explained below:

1. Miyadaiku (Shrine Carpenters)

Here, the carpenters are engaged in the building and construction of Japanese shrines, as well as temples. They make use of well-detailed joints to construct powerfully structured and highly long-lasting structures. As a result, these buildings are commonly found among the world's longest-surviving wooden structures.

2. Sukiya-daiku (Tea-House Makers)

These breeds of carpenters are more popular for their particularly delicate and aesthetic constructions. Typically, they engage in the construction of treehouses and residential-type structures, such as staircases and window frames.

3. Tateguya (Interior Carpenters)

The tateguya are carpenters who take care of interior finishing work. These people are interior finishing experts who build shōji (Japanese sliding doors). Furthermore, they also create carved, small wooden wall decorations that are known as *Ranma*.

4. **Sashimono-shi** (Furniture-Makers)

These furniture-makers are similar to the tateguya. The only difference is that they create more general-purpose furniture, such as chairs, sofas, cabinets and more.

Although, it is not always the case for a Japanese carpenter to work outside the realms of their specific skill or profession, it does happen. There are carpentry workshops in Japan that will perform two separate skill sets. Mostly, the *miyadaiku* and *sukiya-daiku* are usually the two such professions.

Japanese carpenters are artists, indeed. Their joinery strategies, as well as a deep appreciation of sturdy woods, such as the *hinoki*, make them unique in their skills. As such, their skills are something that has been sharpened and perfected for more than a millennium, and they prove their worth. It is admirable to carefully observe Japanese carpenters in action and to see the final product. To this end, this book and many videos, as well as websites available around the Internet, can help you learn more.

The various processes involved in the selection of wood

How do you get started with Japanese joinery? Are there processes to follow and what are the conditions you must meet in order to select the hinoki for your construction and joinery projects?

Well, you're not far from the destination. Here you are!

Now, to go from the selection of wood to the completion of a project involves a series of working processes. These procedures, each of which requires the highest level of craftsmanship, jointly influence the final outcome of the product. However routine some of these processes may become, the carpenter will still deeply consider and carefully judge and apply the habits acquired through many years of practice.

Here, we'll discuss one crucial woodworking manufacturing process, which is the choice of wood for your project. The choice of timber entails choosing the right wood species for the object to be produced and judging the quality of that wood.

Moreover, the choice of the right wood species, such as hinoki, for a project determines the product's durability, unique features and characteristics. Each of the four distinct carpentry professions has a range of wood species, each of which impacts the economic, aesthetic, technical and symbolic conditions.

Economic conditions are determined by the history and origin of the wood, its availability, quality, and price. Therefore, products meant for daily use are produced from local wood, while sophisticated items, such as utensils for a tea ceremony, are produced from imported or precious indigenous wood.

Technical conditions consider those properties and qualities of wood that make it suitable for the different purposes for which it is required. Some of these properties include weight, moisture content, durability, elasticity and flexibility.

Hinoki – Japanese Cypress

The wood from Japanese Cypress is fragrant as a result of its high oil and resin content. It has a powerful scent that smells like aromas of spicy lemon and sweet, and it produces resinous coniferous notes reminiscent of Pine and Cypress.

Also, the wood is highly expensive and produces exceptional timber as a result of its rich, straight grain and its brilliant rot resistance. It has been used to build castles, as well as palaces, temples, shrines, Noh theatres, and traditional japanese baths (Onsen).

It is also highly sought after as an ornamental tree and can be found in many gardens, parks and shrines. As a beginner in the Japanese joinery profession, you can subscribe to the services of any of the four types of Japanese carpenters to help you source this wood.

Examples of Basic Joinery Process

Before these examples are rolled out, it's important you know that they'll highlight the fundamental processes for basic joinery, helping you to get ready to start your project with ease.

Tongue-and-Groove Joint: This joint allows for wood shrinkage. Cut a groove in the edge of one piece of the wood. Also, cut a tongue on the other wood piece to fit into the groove. Get two pieces of finely textured wood of equal size, length and thickness.

The first step is to make a non-through cut through the two mating workpieces of the tongue-and-groove joint by not allowing the saw blade to go all the way through the wood.

Turn on your table saw and run the wood through to cut the first face. This will create a *rabbet* down the edge to form half the tongue. Move on to the next stage by flipping the workpiece end-for-end. Repeat the cut on the same edge of the opposite face, to produce the finished tongue.

Check the fit of the resulting joint. How? Simply by slipping one workpiece into the other.

Slip the tongue firmly into the groove and clamp up the assembly.

Chapter Summary

- How the Japanese carpentry industry uses Japanese joinery;

- There are four types of Japanese carpenters;

- There are various processes involved in the selection of wood; and

- An example of basic joinery was demonstrated.

All of the above were talked about in this interesting chapter.

It's now time to build a simple square frame as we proceed to chapter four.

Chapter Five: Building a Simple Square or Rectangular Frame

Basically, wood joinery refers to joining pieces of wood, timber or lumber together to create other structures. If you're really interested in gaining woodworking skills, take this time out to learn the major strong types of wood joints shown below. After all, the stronger the joints, the more long-lasting the final products.

The unique feature of skilled woodworking is the ability to build tight wood joints. The edges mix seamlessly thereby making two joined pieces look like a single piece. To successfully create most types of wood joints, you'll need to make precise cuts.

It is often a requirement to build simple timber frames in woodwork. Fortunately, there are many simple joints that can be used to create them. Where frames are used, they are often covered with plywood or other man-made boards. When a frame is used, it adds strength to the product. This makes it easier for a relatively thin board to be used, thereby saving on both material and cost.

Building a Simple Frame Using Interlocking Miter Joint

A miter joint is created when two end pieces are cut on angles and fitted together. It is commonly found in the corners of picture frames as well as the upper ends of some styles of doorway casing (trim).

When considering a quality 90-degree mitered corner, the two pieces of wood are cut on opposite 45-degree angles and joined together. In the same vein, when installing trim, the pieces are joined at the seam and then fastened to the framing material on the wall.

However, if you want to create mitered corners for a freestanding object, like a picture frame, the pieces of wood are joined at the joint, using well-constructed dovetail joints that do not require nails and screws to hold them together permanently to each other. Looking at freestanding woodworking projects, nearly all miter joints need both gluing and extra fasteners. But here, we're not going to use any fastener; just handcrafted joints that will go into each other and lock up for life.

Steps to Building the Frame

Step one: Choose frame size and glass

You can use any size of wood for your frames. What really matters is just your taste or preference. You may choose 2" x 2" softwood planed timber for your frames. Maybe your plan is to create nice large frames of different sizes to cover a big wall. You can purchase a piece of glass to fit or use glass from old frames that we all have stashed in some dark corner of the house.

Also, you may as well use an inexpensive diploma frame glass and some small thick glass from old halogen light. If you want to cut the glass, you can use a tile cutter as it will perform a great job. You need to exercise caution here so that you don't cut yourself. Always use safety gloves while performing any cut.

Remember, it is better to adjust the frame to the glass and not the other way.

Step two: Cut the wooden planks

There are two ways to cut 45-degree angles. One cutting option is the manual way, using a 45-degree miter box with a tenon saw. All you have to do is to ensure the blade of the tenon saw is set well straight and at the correct 45-degree angle before cutting. Always test first on scrap wood.

Step three: Decide which molding bit to use for the frame design

There are many ways that you can achieve decorative molding for the frame. You can use a straight bit, a cove bit, a round corner bit and a 45 degree bit. After setting the bit to the right and desired height, as well as the distance from the table fence, ensure you always practice the first cut on a piece of scrap wood to see if the setting and RPM gives you the best finish you want.

When using a straight bit, you should always do the cuts in a few shallow passes. But don't force the wood on the bit in one go, as it might burn the bit and the wood finishing will be very poor.

Another way is to first cut all the frame parts and then run them on the router table or take the full length of wood you want to use. Perform all the moldings before you cut it into the frame size of your choice. Just choose the option that is easier for you.

Step four: Put the piece together

When you're done cutting the frame parts to size, make a dry assembly to know whether the glass fits well. Get the frame parts ready to be put together, then assemble them together, fixing each joint together and mildly hitting the ends of the joints with a small mallet to drive the joints in very well.

Make sure to keep the frame together down on a flat surface. It will prevent a twist that might destroy all your work. Give the frame a light sanding, working up the grit level. Sand the frame properly and clean it well by dusting off the frame and using dye or paint. You can even leave it natural. More so, if you use old pine effect dye, the finished frame will look awesome.

Step five: Create holes for hanging the frame

The aim of using the washer is just to stop the frame from falling, which is unlikely but won't do any harm. Alternatively, you can just drill a hole in the center of the top frame.

You can use a 16mm bit for a shallow hole to fit the washer and 9mm in the inner hole. Lastly, cut a 2mm back support for the glass and close with small pins under each corner of the glass.

Lastly, it is important to highlight here that the miter, which is only slightly stronger than the butt, is used almost exclusively for appearance sake (through burnishing) as the joint hides the exposed end grain of both pieces of wood. This is the standard type for picture frames and small decorative finishes.

With thin wooden boards, the miters can be cut using a handsaw although using a power circular saw with a guide or jig will translate into a more precise cut. However, with a top-quality woodworking project like this one, the methods of joinery are often entirely invisible.

Chapter Summary

This chapter spokemuch on tools and joints. We considered the steps needed to create a simple square frame, how to make joints using proportionate measurements, and the importance of burnishing to create a beautiful product. Now, it's time to move ahead with a bit of an advanced project, such as building a step stool.

Chapter Six:
Construct a Step Stool

The step stool is simple furniture, but exceptionally multi-functional in nature. A step stool is a useful piece of furniture in homes, warehouses, shops, workshops, libraries, offices, and other working environments where it is used for convenient access and maintenance jobs. It can help in accomplishing jobs that need to be done at an elevated position.

In this chapter, come along as you'll be shown how to build a step stool using Japanese joinery. This woodworking project is suitable for a novice who is looking for a task requiring hand-cut joinery. Using simple knowledge you've acquired so far about Japanese joinery, this timeless and functional piece of furniture will be a woodworking project that you'll truly enjoy building.

Step one: Get materials for your step stool

Look for a dry piece of construction lumber that has probably been lying idle in your shed, workshop, or house for years. Such woods are usually free, hard, dry, and won't warp after use for some months. Besides, you'll save yourself money, other resources, and effort. But, if you don't have old lumber, you can purchase finished laminated boards.

Dimensions for your boards are the following: 47.24 inches in length, 13 inches in width, and 1 inch in thickness. After making the step stool, the dimensions should be 17.7 inches in width, 8.66 inches in length, and the height of the legs is 11.81 inches, respectively. The thickness should remain 1 inch. These measures are not compulsory to use, you can use them as a guide and not a rule.

When using construction lumber, especially the one you don't know the source, always ensure you check for any metal objects such as nails that can injure you or even harm your tools. Try your best to comply with all the safety information on your power tool and use it with care.

Step two: Rip the lumber before cutting

Ripping of lumber involves tearing the big board into smaller strips using a handheld planer or any other tool of your choice and then putting it back together. This will make the board more straight and it remains so for many years to come.

To get a straight board, rip the lumber to 2.2 inches or any measurement of your choice. More so, you'll have to measure out 5 pieces of 2.2 inches lumber, so that 4 pieces go for board while one piece is for connecting the legs of the stool together.

If you have a table saw, that'll be better. If not, use the handheld circular saw. Although it has a poor guide, you can make a simple guide in a matter of minutes and use it. Get a straight piece of wood and attach it to the bottom of a circular saw with screws. You can do this by drilling 2 holes through the circular saw base.

Measure the distance between the glue wood and blade of the saw using a measuring tape. Remember the distance is supposed to be 2.2 inches and must be equal on both sides of the blade in order to have a straight cut.

Also, check the depth of your saw blade and adjust it. It should be slightly more than the board thickness. This will eliminate the incident of a back kick of circular saw which often happens when the blade is put too low in the wood. Always clamp your wood pieces before working on them.

Step three: Proceed with cutting the board

Now, hold your saw very tight. Gently press the teeth against the board's edge, then cut. Do not stand at the back of the saw when cutting. Instead, stand on the side to ensure that the saw is behind you so that when you lose control and experience a back kick, the saw will not harm you. This is applicable to table saws.

When you're done with the cutting, you'll have 5 pieces of boards as stated earlier.

Step four: Plane the wood

At this stage, get a handheld planer as it might be difficult sourcing a table jointer or thickness planer. Adjust the planning depth and plane down each piece of wood. The best approach is to plane the wider sides first and take note of how many times you plane it. This is important to maintain the same thickness of each wood piece and assist in gluing.

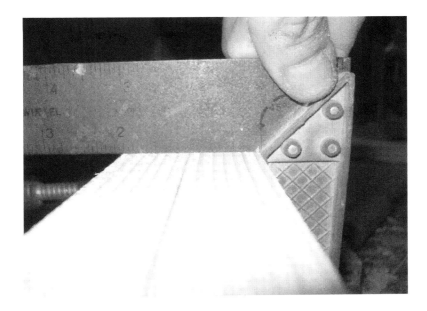

Now, plane the narrow sides, and don't forget to plane them in pairs to keep the pieces squared.

Step five: Put the pieces together

You have to make sure that the wood surfaces are smooth. So, sand narrow parts with 40 grit sandpaper to have an excellent joint. Arrange the pieces uniformly and equally on all sides that will be joined together.

Clamp all the wood pieces together after passing dowel nuts through the narrow parts to make them align well and form the board. Remove excess protrusion from the joints.

Specifically for the fifth wood piece that will connect the legs of the stool together, cut it to 13.77 inches using a regular hand saw or even any other type of saw. Then, leave the clamped piece to dry.

Step six: Flatten and sand the piece

As soon as you remove the clamp, you should have a board that is strong, nice, and elegant. If not so, maybe the board is not flat, especially the edges. So, what will you do?

Put a flat piece of wood and draw a straight line along with that particular flat wood. Remove the wood and stop when you reach the line. Use anything that is available to you to sand it out and repeat it at every end of the board that requires to be flattened. But, don't remove too much wood.

In the same manner, flatten the top and bottom sides of the board as you did to the ends of the board earlier. You can use a hand plane to achieve that and use a piece of straight scrap wood to check whether the sides are flat.

If eventually, you made the mistake of creating a dent in the wood while planning it, use wood filler to fix it. The next step is to sand the product which you can do effectively using any kind of sandpaper or hand sanding

Sandpaper may not save a lot of time, but it produces a better effect. If you're sanding by hand, use either sanding block or some pieces of hardwood between sandpaper and your hand.

Step seven: Cut the wood

Since your board is now flat and rough sanded, it's time to cut it to the pieces needed for the stool. You'll have to cut a total of three pieces, 2 for the leg of the stool and 1 for the top.

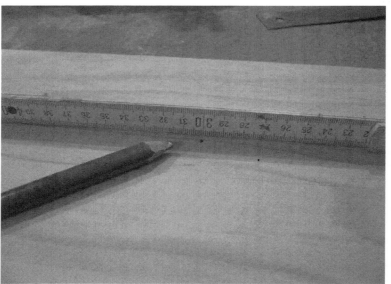

Use a hand saw or any other device for cutting. Cut the board on one side to make a square and remove glued pieces that are unequal finish. Check for square and move on with marking. The next step is to mark the legs and top board with 11.8 and 16.5 inches respectively, then cut through.

Step eight: Draw a layout and create a dado joint

The legs of the stool should not be a block of wood. So, you need to draw a layout for them. Get a caliper or divider or even a round bucket cover and make a circle of any arbitrary size.

Measure out the bottom side of the legs and add markings on both sides. You can use fewer circles for the bottom pattern. Remember, these dimensions are just guides, you're free to use other circle patterns.

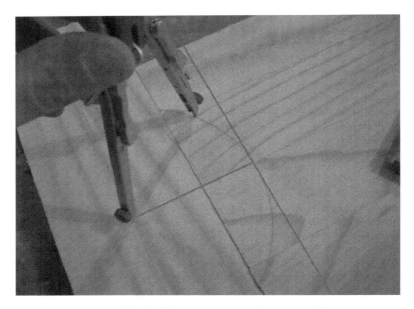

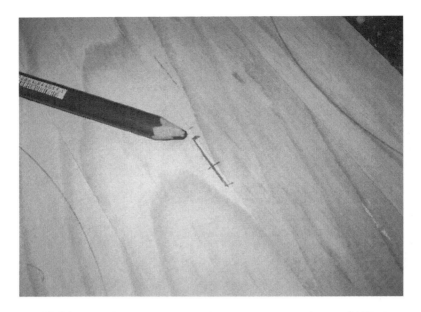

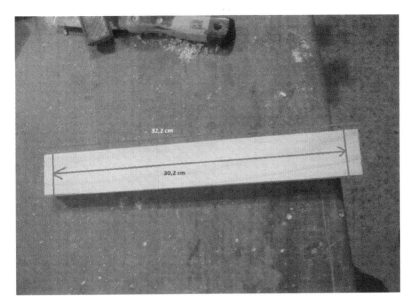

The next step is to make a dado joint – this stage is a bit technical and requires your patience and carefulness. Make a female slot on the bottom side of the top board. Put 2 inches of the legs inside the board and 0.2 inches from the edge of the board.

Look for a router as it is the best tool at this stage. Use 0.5 inches straight bit and rout 0.6 inches deep. To rout the female joint, use flat guide wood. After these measurements, mark the distance between the guide wood and the edge of the board and note it. When you want to rout the other side of the board, just put the guide wood at the same distance from the edge of the board. That's all for the female part of the joint.

Now, rout out the male part of the joint on the stool's legs. Make the male parts of the joint 0.55 inches wide and 0.5 inches thick to correspond to the width of the female slot. The male part of the joint is to be made on both legs as well as on the side that will attach to the top board.

Having determined and made these markings on the wood. Clamp your work very tight using guide wood just as you did before. Rout the two legs simultaneously and cut 2 centimeters of the male joint from both sides of the legs using a chisel and handsaw. Your joint should fit snugly. But don't join the legs to the board yet.

Prepare the connection piece between the two legs by first of all making female slots of the legs of that piece. Again, dado joints are needed here. You'll need to make slots on the inside of the legs. Locate the center of the leg and mark the slot vertically, making it a little smaller than the connection piece. Use guide wood to rout the slot on the two legs.

Measure the space between both legs to make the male part of the joint on the connection piece. Cut the connection piece slightly bigger for the male slot on each side. Rout the wood to make a male piece of joint, use a hand saw to remove 0.2 inches from the male part of the joint to fit it to the slot.

Step nine: Cut the layout on the legs and sand it

Do you remember the markings you made earlier? Now is the time to use them. You can use a hand saw for the cutting. Cut by the line and not into the line. After cutting, sand the surfaces to smoothen them using sandpaper for edge sanding.

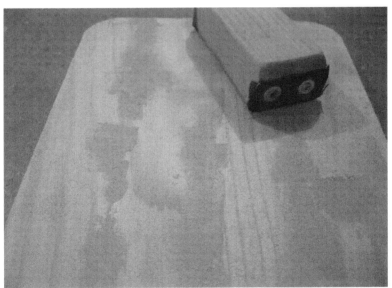

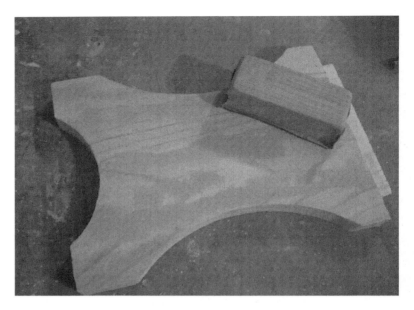

If you curve the stool's leg, apply wood filler where necessary and allow it to dry. Then, sand those surfaces again. You know that a sanding belt is not required here, therefore use a sanding block and a piece of either 120 or 150 grit sandpaper.

Then, fine-tune or profile the edges. To profile the parts very well, use a router and round router bit or opt for a file and sandpaper. Just insert the right bit into the router and profile the edges. You may sand the entire parts again to ensure there are no rough surfaces on it.

Step ten: Put the piece together

Since you're using dado joints, your joints require to be put together and they'll last for many years. So, kindly put the parts together and use your big mallet to drive them in. If you care, you can check whether the legs of the stool are properly aligned and flat. Just place the stool on a very flat surface and if it teeters, check for the leg that is longer and sand it accordingly.

Chapter Summary

Now that you've learned how to make a live wooden object, you can begin to visualize the maximum possibilities of creating an advanced product. In the next chapter, you'll learn about a slightly more complex project of making a tool chest. Come along as the ride promises to be interesting and enjoyable.

Chapter Seven: Create A Tool Chest

A tool chest is a great way to keep your tools organized at home or in your workshop.

It is more than just a place you can keep your tools. It's an essential part of your toolset no matter your level of proficiency. In other words, whether it's DIY or a professional car garage, a good tool chest is an absolute need. Bearing all this in mind, you should make every effort to build a tool chest that is strong, reliable, and efficient.

Why Do You Need A Tool Chest?

Here are the reasons a tool chest or toolbox is such an essential thing to have;

- Your tools are in a convenient place, ready for you whenever you need them

- They're also safely and securely stored away

- An organized tool chest can make your tasks easier, faster, and more effective

- Having tools in a box or chest keeps them protected and reduces the risk of damage

A tool chest, as the name sounds, is similar to a chest of draws. The tool chest is bigger in size and their largest design feature is to store as many tools as needed. You'll typically find tool chests in the home or professional garages. They are stationaries, although some may have wheels to enable easy transportation. Their size and weight are not the best design if you regularly need to take your tools from job to job.

So, how do you make one based on your knowledge of Japanese carpentry and joinery? Well, that's very simple. Here are the steps to follow.

Step one: Choose the right dimension and cut the wood

Deciding whether you want your tool chest to be big, small, or of average size is the basic step to follow. Of course, your choice depends on the number of tools you would want your box to carry. Generally speaking, you want it to house even your longest tool – the saw.

For this project, use 1x10 inch pine wood and cut the pieces to size. Use the following dimensions:

A. 1 @ 15 x 9.25 inches (bottom)

B. 2 @ 15 x 6 inches (long sides)

C. 2 @ 9.25 x 5.25 inches (short sides)

D. 2 @ 10.75 x 2.5 inches (top sides)

E. 1 @ 10.75 x 1.5 inches (lid side)

F. 1 @ 10.75 x 2.5 inches (wedge & wedge fitting, cut in the middle at an angle)

G. 1 @ 9.25 x 11 inches (lid)

H. 2 @ 2 x 9.25 inches (handles)

I. 2 @ 4.5 X 2.25 inches (feet)

Step two: Get the handles and wedge ready

Simply cut a curve for the handles using the hand saw and chisel. In the same vein, you should use the hand saw and chisel to cut the wedge at an angle. See the pictures below:

Step three: Prepare the handtools

The hand plane can assist you in cleaning up the angled wedge while the spokeshave does the work of smoothing out the inside of the handles.

Step four: Mark and create holes

At this point, you need to mark out the spots where all the plug holes would go to hold the box firmly tight. After marking these spots, begin to create these holes. These holes are necessary as they add to the beauty of the box and make it strong eventually.

Step five: Assemble the parts together

Once the holes are drilled, it's time to assemble the pieces together with the help of screws. Traditionally, the method of joining Japanese tool chest is through wooden plugs as they are easier to use and they offer the opportunity for plugs to work well.

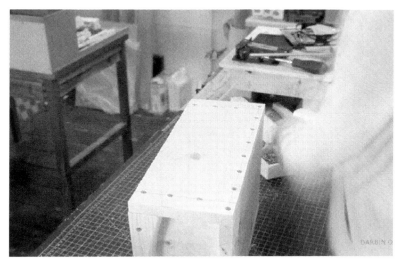

Step Six: Add legs to the base

If you wish to slightly bring the tool chest up from the ground, then you can add about 0.25 inches to the product.

However, this step is an optional one. You can ignore it or implement it if you want to.

Step seven: Create a space for the handle

To make the tool chest very easy to carry and lift the cover whenever the need arises, you will have to drill two holes in the cover (lid). These holes should be large enough for some ropes to pass through and fit in.

Step Eight: Sand the box

Now, you need to sand every part of the box to make it look nice, smooth, and beautiful. Just use your sandpaper to obtain uniform sanding of all the surfaces.

Step Nine: Stain the tool chest

This is yet another optional step. You may decide to either stain some parts or the whole box, or you may wish to leave it natural, the way it is. The choice is yours.

Step ten: Insert the plugs

At this point in time, you need to fill in the holes with screws. So, measure the depth of the countersink. Then, set up a stop block on the bandsaw to cut the required plugs. Apply a bead of glue on each plug and use a mallet to securely push them inside the holes.

Step Eleven: Sand those spots

To make the inserted plugs look nice, neat, and smooth call your local sanding machine (sandpaper) back to work. Use the sandpaper to sand all those spots where you inserted the plugs to smoothen them. You can even apply stain to those areas if you wish to make them look uniform.

Step Twelve: Fix the rope

At this last step, you have to add the rope for the handle. All you'll have to do is to simply tie two knots on the underside of the cover (lid) to securely keep the handle in place.

For a final finish on the box, you may decide to put a touch of shellac all over the outside of the box to beautify and give it some protection.

That's all for this classic Japanese tool chest.

Chapter Summary

The above project was an interesting one in the sense that it took us to the practical class of building a tool chest from scratch using the Japanese joinery concept. With this tool chest, your tools will now have a place to stay for easy location. If your tools are organized, your work will be organized too. This project is pivotal to your success as a craftsman!

Chapter Eight:
Make A Dynamic And Beautiful Dinner Table

You've tested your hands on small projects and become familiar with Japanese joinery. The skill gained from these several projects will make it easier for you to handle tools and other materials. Based on this, you can equally handle large projects. It's now time to create a bigger project – a beautiful dining table.

By the way, how should a dining table be with regards to space, size, and accommodation?

Your first step in choosing the perfect dining table is to consider the space of the room it will go in. Is the space for a formal dining room you use to host dinner parties? Or will the room be used for everyday settings for homework and most meals? For an everyday-use arrangement, a dining table that is low maintenance but the high style may be your best bet. Make sure that whatever you decide on, you're leaving enough space surrounding the table to comfortably walk around and get in and out of the seats.

Understanding the size and material you're looking for maybe the most difficult part. From a traditional setting to a more modern style, your dining table can do more or less as you want it to. Therefore, you need to keep it classic with an all-wood option. You can even get loud and select something that will set the pace for style in your dining room. Combined with the most suitable chairs, your dining table should display your lifestyle at its best.

The shape of the table plays an important role in creating an elegant and comfortable space. Presented below is a guide to help you determine what size of dining table you may need once you've chosen the right shape.

Size	Oval & Rectangular	Round & Square
48 inches	4 persons	4 persons
60 inches	6 persons	6 persons
72 inches	6 persons	6 persons
96 inches	8 persons	Not a typical size
120 inches	10 persons	Not a typical size

Understand the fact that every person requires about 24 inches of eating space. Besides, your table should be at least 36 inches wide to give enough room for food and place settings.

Construction of traditional Japanese dining table

To get started, two beams are secured to the tabletop with sliding dovetails. The legs are set and fixed into that beam with mortise-and-tenon joints. There is neither an apron nor a brace between legs. As such, there's no leg-to-tabletop linkage. Here is the working procedure for this table. First start with the tabletop, work on the sliding dovetail beam and conclude with the legs.

Sketch and Design of the Table

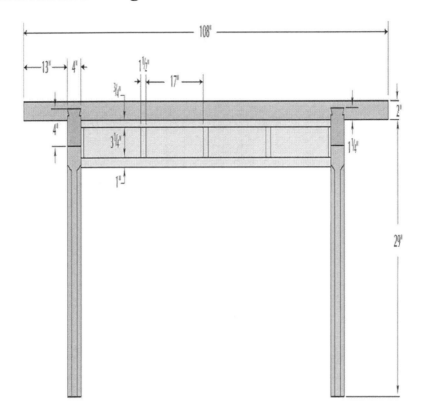

Elevation

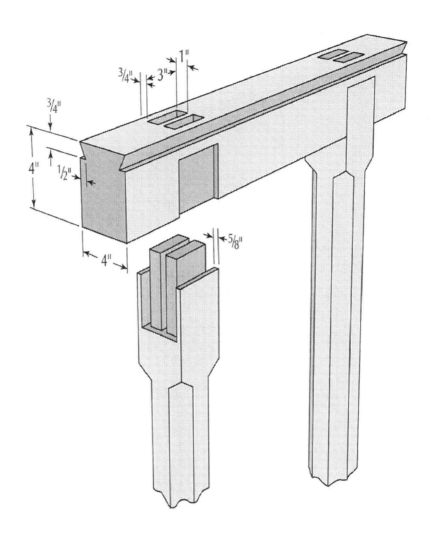

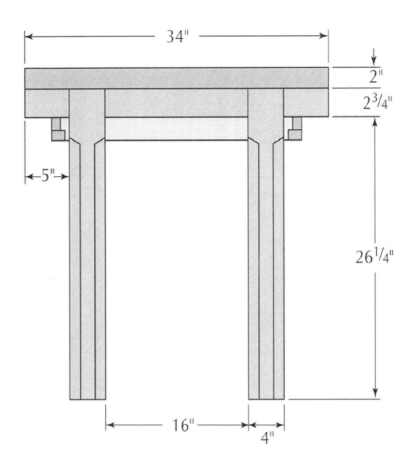

Profile

Dimensions:

Item	Number	Sizes (inches) Thickness Length Width		
Tabletop	1	2	108	34
Legs	4	4	36	4
Beams	2	4	36	4

Joints that'll make the construction feasible

You need to mark the position of the legs of the table. So, mark the edge of the table all around the beams and decide on the position of the legs. Since the table does not have an apron or leg brace, it means that all the stress retires to this point. You need to design a joint that'll absorb all the stress coming from all angles.

Use double mortise-and-tenon joints and place a non-shouldered tenon on both ends to be in a small space with the tenon joints. The non-shouldered tenon is on the same level as the beam. This ensures that the table is secure in both length and width.

This you can achieve by pounding the beams out and mark the mortise positions. Roughly excavate the waste from the two sides. Then, clean the mortises from the two sides using a Japanese chisel.

You'll need to cut out about 0.63 inches for the non-shouldered tenons that offer additional support to the legs and remove the material to have a smooth surface.

Here are the steps:

Step one: Get your wood pieces

You need to figure out how much wood you need for this project and go pick it up. Maybe you have a log of wood (ash) you acquired over the years that has been lying idle in your workshop, this is the time to use it. Or you can look for straight boards you can find and use. All of the wood needed for this project won't cost you much, less than $100 or thereabout.

Step two: Cut the wood

It's possible to get the guys at the hardware shop to do the cutting for you, but the ideal thing is to cut the pieces by yourself. Mark out the dimensions and cut using a hand saw. Make sure you label each piece of wood you cut so that you don't get confused later.

Step three: Plane the tabletop

At this stage, you have to flatten the two sides of the tabletop using sandpaper. This will carefully remove twist and warp and also keep the top at an optimum thickness. Remember to plane across the grain of the wood.

Also, you can conveniently use 250 grit sandpaper to smoothen the surface of the tabletop.

Step four: Set up the dovetail beam

The dovetail beam is the next item on the list. Considering the fact that the tabletop is about 2 inches thick, then the legs and beam should be about 4 x 4 x 36, all in inches taking into account the part of the beam that enters into the tabletop with a tail.

All you have to do is to use the jointer and band saw to create all the pieces. Then, mark the position of the sliding dovetail beams by taking a look at the movement of the wood's grain and color. Also, the four legs of the table were selected in this order – the left and right first, followed by the front and backside.

Go to the bottom side of the tabletop and mark the centerline from one end to the other end with an ink line. Mark the table ends squarely on the two sides of the table from the centerline. The positions of the sliding dovetail beam are marked from the end lines.

Step five: Create a strong connection

Considering the weight of the table, the design has no connection for its legs and apron under the table. So, depending on just the tail to make the legs strong is not ideal. The best option is to sink part of the beam into the tabletop together with its tail. This will provide more than enough strength in the brace.

It is important you consider the size and depth of the tail support as well as decide on the tail's angle before marking out the lines. You can use a narrow Japanese plane, circular saw, and chisels to remove the waste. Use a straightedge to check the flatness so as to create the tapered sliding dovetail groove.

Step six: Measure the beam's tapered sliding dovetail

With a knife and marking gauge, mark the beam's tapered sliding dovetail and use the table saw to roughly cut out the tail.

Carefully plane both tails using a planer and chisel. Test the fit while you work. Finish the pins and tails of the joint and cut a large chamfer on the two ends of the tail beam. Then, pound the beam into the groove using a big mallet.

Step seven: Cut the end of the legs

Test your skills here by cutting the end of the legs cleanly and squarely. Mark the tenons and rip the leg's tenon with a Japanese handheld ripsaw. Use a chisel to remove the materials in between.

Step eight: Draw the octagonal layout of the leg

Octagon is the best shape for the leg. From the hard cardboard, make an octagonal template and trace it to the bottom end of the legs. Draw the line up the face of the leg until it is 10.5 inches below the beamline using a marking gauge. The octagon layout determines the chamfered corners, while the line set with the marking gauge determines the final surface. Draw lamb's tongues to not disturb the tenons.

Step nine: Cut the leg corners

Kindly cut out the leg corners marked out above with a hand saw. Also, ensure you fix your eyes on the edges to cut on the marked lines without making any mistakes and cut the leg corners up to the lamb's tongues.

Step ten: Smoothen the final surface

Cutting the legs will give it a rough octagonal shape. Use a drawknife, Japanese hand plane, and chisels to cleanly smoothen the octagonal-shaped legs and lamb's tongue.

Step eleven: Insert the legs

This is the table smackdown stage! The end of the legs is chamfered to receive the big mallet.

Step twelve: Cut a little slit for the wedges

Assemble the mortise and tenon joints with glue and wedges. You should cut a small slit for wedges on the tenon, but remember that the non-shouldered tenons are on the two sides of the center tenons. This makes it hard to use a saw to rip the slits. However, the outside tenons stop at the bottom of the tabletop. So, the center tenons will be longer than the outside tenons. Cut the outside tenons to their actual length and rip the slits with a small Japanese rip saw. Ensure that all the legs are fitted and the joints are adjusted accordingly.

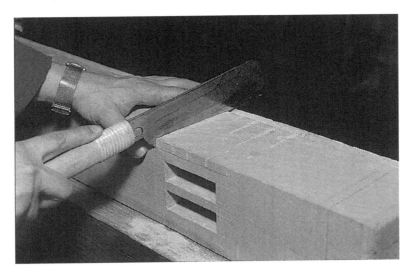

Step thirteen: Complete the leg-to-beam joint

After cutting and planning every part, use a plane sander to smoothen all the legs, the bottom surface of the table, and sliding dovetail beams. Now, you're ready to assemble the legs to the sliding beam using glue. Pound one leg at a time so that the tenons will come out of the beam, fitting tightly.

Hammer and glue the wedges into the slits and leave the leg there to assemble another one. Return to the first leg to cut and plane the tenons flush with the beams. The other leg is finished likewise.

Clean and sand the non-shouldered tenons with a palm sander. Plane the top of the leg's tenon before the final assembly so that the tenons will not touch the tabletop in the future. You can pound the beam into the top for the last time.

Step fourteen: Attach the legs and beams

The end of the beams together with the attached legs is pounded while the legs gradually moved to the center of the table. Finally, the beam mark will eventually come to the edge of the table. You can now measure the length of the legs evenly and cut them with a Japanese saw. The octagon edges should be chamfered and the end grain planted. That's all!

Chapter Summary

One of the most practical chapters of this book—chapter seven—took into consideration the following important subjects:

- The importance of a dining room vis-à-vis a dining table in your house.

- The influence of size, shape, style, and space in choosing a dining table.

- Design of the table on paper.

- What joints would make the construction worthwhile?

- The step-by-step procedures for building a dining table.

It's time to look at another concept—Lacquering and preservation of wooden objects.

It deals with the meaning, kinds of Japanese lacquering, etc. Chapter 8 promises to be interesting and insightful. So, let's go there!

Chapter Nine: Lacquering and Preserving a Wooden Object

Lacquer is a term used to refer to a range of hard and shiny finishes which are applied to woods or metals. In modern techniques, it means a number of pigmented or clear coatings that dry through the process of solvent evaporation to yield a sturdy but durable finish.

There is the Japanese lacquerware which is also known as true lacquer. They're objects coated with the treated, dried, and dyed sap of a tree called *Toxicodendron vernicifluum* (*the Japanese varnish tree*) or any other related tree. This Japanese lacquer – also known as **Urushi** is applied in many coats to a base that is usually wood. It dries to a very hard and smooth layer which is strong, beautiful, long-lasting, waterproof, and attractive to the eyes.

Permit me to explain here that lacquerware is the art of building art works and designs using the sap of the Urushi tree. The sap on its own contains one powerful resin known as urushiol. This resin polymerizes under air and moisture to become a hard, durable, and plastic-like substance called Japanese lacquer.

When it dries, Japanese lacquer stops the attack of moisture and forms a sturdy film that precludes decay. Little wonder lacquer has been applied on everyday items right from ancient days. In japan, lacquerware items such as trays, bowls, chopsticks, and many-tiered boxes are easily found there.

They're sometimes painted with pictures, carved, dusted with gold, inlaid with other materials, as well as given other decorative treatments. The techniques used in lacquer have been perfected over many centuries ago so as to produce beautifully detailed designs for furniture, woods, and boxes meant for domestic use.

Kinds of Japanese Lacquering

Japanese lacquer differs in quality, color, and surface texture. But lacquer has pigments added to it and traditional pigment colors include black, red, and off-white, green, and yellow.

The following are the various types of Japanese lacquer (Urushi).

Tame Nuri: It makes use of a clear lacquer that is applied on base material. This allows the elegance of the base material to be very visible.

Tsumakure: It has red accents painted on items. This type of Japanese lacquer is often seen on the edges of *tana*, which literally means *red fingernails*.

Shin Nuri: This type has a smooth black shiny layer. Multiple coats of lacquer are applied to the wood, sanded, and smoothened to achieve this type of lacquer.

Tatakinuri: It has a non-shiny but rough texture. This 'pebbly' texture is achieved by mixing different materials into the lacquer including tofu, crushed eggshells, and okara. Okara are small pieces of materials left over in the manufacturing process of tofu. Now, the mixture is applied to the object using a sponge and tapping (tataki) motion. The surface is smoothed using a roller. As a result, this Japanese lacquer type creates a hard surface. It has been applied to the Japanese body armor.

Kaki-awase Nuri: this is the lowest grade of black lacquer. A coat of seshime-urushi mixed with lamp-black is applied to the wood and it hardens the surface of the wood as well as stains it with black color. This is followed by one good coat of joohana-urushi or joochin-urushi.

How Wooden Objects are Prone to Fading, Termite and Fungal Attack

Wooden objects are prone to fading, termite and fungal attack as well as other problems based on the kind of wood involved. The truth is the wooden products are subject to deterioration from a number of biological agents such as fungi, termites; physical deterioration such as fading, poor handling, as well as other issues ranging from heat to light and chemical degradation. Let's discuss them one after the other, in three stages.

- **Biological Degradation**

Given a favorable condition, a variety of different organisms will attack your wooden object. They include termites, fungi, bacteria, etc. For instance, termites such as furniture beetles, white and black ants as well as other insects cause serious damage to wooden objects especially softwoods like pines and fir, old furniture made of oak and walnut, and hardwoods.

They're also found in museums, damp timber or in structural timbers of recent buildings. Most of these organisms thrive well in the presence of moisture, limited amount of air, and moderate temperatures.

The sapwood and heartwood of most wooden species are prone to attack by a range of various wood rotting and staining fungi. Most of them flourish in dirty, damp, and unventilated environments.

- **Physical Degradation**

Wooden objects may be physically damaged by either poor handling or as a result of stresses induced by changing moisture gradients. Its relative softness means that the object can be easily damaged by contact with harder and sharper objects or surfaces.

Excessively hot and dry conditions cause wood objects to shrink and crack. Conversely, cold and damp conditions cause these objects to swell and warp. Substantial damage may be caused by large, rapid fluctuations in relative humidity levels. This is particularly so if wooden parts are closely joined and if their respective grains run contrary to one another. The restriction of free movement often leads to warping and cracking.

The rate of change of relative humidity levels and the way in which the wood has been sawn (along or across the grain, radially) determine whether bowed, cupped, twisted, cracked or split wood will result from exposure to inappropriate or changing relative humidity conditions.

The expansion and contraction of wood may damage either the wood itself or materials attached to it. Physical damage may occur if no precautions are taken when wooden objects are transported from one climatic region to another. For example, wooden sculptures transported from tropical regions to drier areas commonly develop serious cracks as moisture is lost from the wood. Similar damage may occur if items composed of wood or other organic substances are exposed to hot sunlight during the day and cold, moist conditions at night.

- **Other Issues**

Included in this category are thermal (heat), chemical, and light. The most drastic form of thermal degradation is the complete damage of wood by heat, fire, burning. Physical damage will happen if a wooden object is placed close to a heat source. The resultant loss of moisture often brings about shrinking and cracking of the wood. Thermal degradation is considered to be the most deadly threat to wood items kept indoors.

Chemical, mechanical and light energy factors combine to contribute to the deterioration of wooden objects that are exposed outdoors (weathering). The general appearance and surface finishes of historic structures are often affected.

When wood is kept reasonably dry and exposed to sunlight or UV radiation, the surface tends to turn brown. A grey finish is observed when the effects of light and moisture are combined. Excessive light exposure will cause bleaching of certain dyes or pigments and fading or discoloration of surface finishes.

Preservation Techniques

Rot-inducing fungi can be stopped by removing one of these four elements that make the fungi to live:

- Enough moisture

- Oxygen

- Food

- Temperature

Only one of these elements needs to be removed to prevent wood decay. The easiest method is to keep the wooden item dry. Most rot-causing fungi will not attack wood if the moisture content is less than 20 percent.

The oxygen component can be removed by submerging the wood in water. Logs that cannot be processed soon after felling and bucking should be placed under a water sprinkler system or submerged in water.

The food component cannot be removed, but it could be poisoned by preservative-treating the wood. Also, temperatures below 50 degrees Fahrenheit will allow negligible fungi growth, and temperatures above 200 degrees Fahrenheit are deadly to fungi.

Handling Decay

Since most decay problems are caused by moisture, the solution is quite simple. Eliminate the source of moisture. Check the roof, walls and plumbing for leaks. Go outside and check the eaves and gutters.

Are the eaves wide enough to prevent water from coming down the sidewalls? Are your gutters poorly maintained or missing? Be sure the foundation is not cracked and the soil slopes away from the house. Don't just treat the mildew, mold, or decay.

If the decay is too drastic, and you want to preserve the historic or architectural character of moldings, carvings, or furniture, consider an epoxy repair job. Epoxies contain resin and hardeners that are mixed just before use.

Liquids for injection and spatula-applied pastes are available. After curing, epoxy-stabilized wood can be shaped with regular woodworking tools and painted. Epoxies are no preservatives and will not stop existing decay. They can be tricky to use, so follow all label directions.

The prevention and control of termites are based on the same factors that affect the growth of wood-destroying fungi as stated above. The best method to prevent attack by termites is to build wooden structures in a manner that allows the wood to be kept dry.

Structural and sanitary measures will not give complete protection against termites, so a chemical means of protection is often the best. The ideal time to install a chemical barrier under a home or shed is at the time of construction. But if a building becomes infested, steps can be taken to dry the wood or construct a chemical barrier between the nest and the infested wooden object.

Also, nests and potential nesting places near the building should be eliminated if they can be found. Contact a reputable pest control company for assistance when termites attack wooden structures.

Preservative Treatments

Factory-applied preservatives fall into two general classes: those with an oily nature, such as creosote and petroleum solutions of pentachlorophenol, and those that are dissolved or suspended in water and applied as water solutions. The main difference is the type of liquid used to carry the toxic chemicals into the wood structure.

Heavy oil preservatives have some advantage in extremely wet situations, since besides being toxic to fungi, the liquid carrier slows liquid water movement. A serious drawback to the oil-based treatments is that the wood surface is oily and difficult to finish or paint.

It is possible to use light organic solvents as the carrier for toxic compounds so the wood may be painted after treatment. These solvents evaporate rapidly, leaving the wood with an untreated appearance.

Preventive Techniques / Conservation

In order to ensure your wooden objects are well-taken care of, take the following factors into account:

- Light, temperature, and relative humidity;
- Handling techniques;
- Modes of storage, display and support;
- Protection from insects, fungi, and dirt.

Environmental Conditions

To maintain wooden objects in the best condition, the following environmental conditions are ideal and suitable:

- The relative humidity levels should be in the range of 40 to 60 percent, with a maximum variation of 5 percent in any 24 hour period.

- The temperature range of 15 to 25 degrees Centigrade with a maximum variation of 4 degrees Centigrade in any 24 hour period.

- Also, the light levels of 50 lux for dyed or painted wood, up to 200 lux for undyed or uncoated wood and a maximum of 300 lux for wooden objects that have largely been used outdoors or have otherwise lost their natural coloring.

- Do not keep wooden objects in direct contact with outside walls or in areas in which large variations in temperature and relative humidity are expected. Avoid keeping wooden objects near fireplaces, heaters, air conditioning vents and doorways. Maintaining relative humidity levels below 65 percent should ensure that fungal attack does not occur.

- In a situation where both metal and wooden components are present in a wooden object, you need to favor one over the other. In these cases, it is preferable to make the conditions more favorable to the wood. This is so because wood is more sensitive to changes in environmental moisture levels than metals.

- Do not expose furniture and other wooden artefacts to direct sunlight. In addition to causing photochemical damage to the wood itself, the joints might be affected resulting in lifting, shrinkage, warping and cracking.

Handling the Objects

Common sense is the best guide when moving or handling any object. Follow the guidelines below:

- Look for assistance when moving large pieces of furniture.

- Always plan ahead. Clear the pathway to and the final location for the object. Ensure you move objects in a slow manner to avoid a fall.

- Hold objects tightly only in areas that can support their full weight such as the rail of a chair and the apron of a table (not the legs or tabletop).

- Remove any detachable pieces before movement.

- Do not drag the furniture on the floor when moving it. If not, the side thrust on feet or legs can place undue pressure on joints.

- Do not use gloves so as to minimize the risk of dropping wooden objects.

Storage and Display of the Objects

The following guidelines will help you to properly store, showcase, and support wooden objects safely:

- Use stable, inert materials for the construction and support of artefacts.

- The size of the objects themselves determines the mode of storage. So, drawers, shelves, cupboards or even the floor itself may be right.

- If drawers are used for storage, use enameled metal drawers instead. You can use wooden drawers, but run away from chipboard as well as other composition boards.

- Store flat wooden objects on level surfaces. For objects having irregular surfaces, look for specially constructed supports or padding.

- Padding may be used only if the object's stability is not compromised.

- Big flat-bottomed objects may be stored on the floor but should be raised on padded blocks to allow for even air circulation.

- Use dust covers for covering and keep furniture clean and free from dust.

- Do not see historic furniture as ordinary furniture as they form part of a collection. For instance, do not sit on the chairs in such a collection.

- Be careful what is placed on a piece of furniture as sharp objects may scratch the surface and hot items, condensation or liquid spills can badly affect surface finishes.

- Maintain a stable, clean environment.

- Put bubble wrap and dust covers over large wooden objects.

- Do not keep or consume food and drink very close to the wooden artefacts.

- Inspect objects regularly, looking for signs of insect attack such as flight holes that may have fallen from such holes.

- Large wooden objects kept outdoors should be kept under cover on a concrete pad to protect them from weathering elements. This will help to stop undue access by black and white ants. Extra protection from weathering can be achieved by maintaining painted or varnished surfaces. Dust these objects regularly.

- Use stands to raise wheeled objects from the ground. This takes the weight of the vehicle from the wheels and reduces access by ants.

- If transporting wooden objects between regions of differing relative humidity, take precautions to allow the object to acclimatize to its new environment.

Chapter Summary

This chapter discussed everything you need to know about Japanese lacquering. It started with introduction to the concept and delve into many other areas such as:

- Kinds of Japanese lacquering

- How wooden objects are prone to fading, fungi attack, etc.

- The preventive techniques you can take to stop it.

The next item on the list is a ride into Japanese joinery and Taoism—the Zen philosophy. We'll look at the relationship between Japanese joinery and Taoism and how they connect to Japanese spiritual culture. If you're ready, let's go!

Chapter Ten:
Japanese Joinery And Taoism

For many people, the attraction of Japanese joinery is dependent on the fact that it reminds them of the very old world of handcrafts. Though not existing now, it was an era that was in peace with nature through the careful observation of its cycles and rhythms.

In the same vein, it is the aesthetic of the assembly for others, whose technique comes precisely from that observation of the natural rhythms presented by the geometric union, often invisible, of all the parts. The process of interlocking that joins the wood through self-sustaining joints was already an extensively used skill. This skill was so much around when the *Miya-Daiku* built their famous Zen temples and teahouses.

Japanese carpentry keeps the singularity of each one of its master craftsmen. It also conserves the spirit of the tree that produced each piece. But all of the changes that happen in the creative process belong to a subtle and discreet world. This is similar to the hidden joints in furniture or houses where what matters is the beauty, strength, and long-lasting quality of the wood used in the construction process.

The invisible assembly seen in the modern world represents the survival of an art form that directly imitates the discretion of nature. It is not the interest in the piece of furniture itself that is important, but what is more important is a good piece that appears to be meditating in place. As you know, a master craftsman would hardly design a plastic chair.

Japanese joinery and woodworking is seen as something that is far more than simply a trade. Many see it as an artform that includes Japanese philosophical knowledge of aesthetics to build strong and excellent products. Japanese craftsmanship is founded on the skill developed to a high degree to which Japanese woodworking is the highest example.

One of the reasons Japanese woodworking and joinery are still relevant and widely used is as a result of the demand for handmade superlative furniture which has survived the industrialization of the wood and furniture industry.

Also, one of the most attractive and interesting parts of Japanese woodworking is that every joint is compressively held together without screws and glue. The joints are precisely crafted to enable the tightness of the joint in such a way that it doesn't warp or break. Typically, the wood from naturally fallen trees is used to form the foundation of the structure or object to be made.

In the world of traditional Japanese woodworking and joinery, it is uncommon to introduce new methods and technologies within the craft. The introduction of new technologies makes the creative processes lose the human touch as well as the wood. Rather, most artisans aim to hold fast the historical methods and techniques within the Japanese joinery art.

To fully understand how Japanese joinery developed alongside its deeply rooted cultural beliefs, one must look back into the history of Japanese Taoism and how it connects to Japanese joinery.

Aesthetic Zen Philosophy

Majority of the values seen in the way Japanese have chosen to create and decorate furniture is traceable to two important religions namely; native Shinto and Chinese-controlled Zen Buddhism, with a particular reference to the tea ceremony.

Shintoism

The most important belief of Shintoism is the worship of kami, the spirits that live in people, places, inanimate objects like trees, stones and other natural existing things. The worship for nature has deep influence on Japanese craftsmen who want to offer trees and even their kami a new life by way of creating nice objects with the wood.

So the respect Japanese have for trees makes them respect the wood. Lumber is cut for the sole purpose of reducing wastage and maximizing the elegance and character of each wood. Large layers are finished to a mirror polish. Many wood pieces are finished in a way to mark the beauty and grain of the wood, either by using a clear lacquer or leaving the wood unfinished.

Strong joinery is applied to create lasting wooden objects that will preserve the spirit of the tree for a long time. Little wonder, this same reverence for the tree is influenced by Japanese way of not tampering with the natural edge of the boards, whether on the edge of the table or on the surfaces of architectural beams and posts.

Zen, Tea Ceremony, and Discipline

The Zen Buddhism and its associated tea ceremony came from China and quickly became a ritual practice among the Japanese elite. As a result of Japanese long history of taking part in the tea ceremony, the act of deciding how visibly old and decaying things have become a major traditional aesthetic, even outside of the tea house. This results in a love for wooden and furniture objects.

Japanese Zen Buddhism and Chinese Chan Buddhism were attractive to martial groups in the two countries. The tea ceremony reflects this through its styled nature in which every object has its complete form and place. In the same vein, furniture – which are designed to be evocative and mysterious of the spirit of nature, is also built to be absolutely accurate, simple, and highly stylized in use.

The discipline needed in Zen Buddhism supported patience and perfectionism. This translated into furniture that is beautiful and deceptively simple in looking. The complexity, accuracy, and precision with which Japanese joinery is built is a distinguishing characteristic of Japanese furniture.

Chapter Summary

As seen in many other components of Japanese culture and tradition, it is the combination of native Shinto ideas and those of Zen Buddhism that weave the excellent tapestry of Japanese woodworking and furniture traditions. So, Japanese joinery has a strong connection with Taoism and both work to preserve Japanese culture and tradition through woodworking and carpentry.

Final Words

Nowhere is joinery and traditional carpentry more apparent than Japan, a nation with an architectural tradition like no other. Long before screws, nails, and metal fastenings became the order of the day, Japanese craftsmen had already become professionals in the art of wood joinery. Using techniques handed down in guilds and families for centuries, Japanese craftsmen, artisans, woodworkers, and builders would fit wooden beams together without any external fasteners. Buildings and other wooden objects would stand for generations, held together with nothing more than tension and friction.

Currently, japan boasts of many craftsmen actively engaged in the pursuit of traditional woodworking crafts. The good part of it is that many of them are well-advanced in years. But unfortunately, only a few young Japanese show interest in carrying on these age-long traditions of their fathers.

My hope is that this book will leave you, the reader, with a general understanding of the importance of woodworking with a particular reference to Japanese joinery. With this knowledge, you can be able to make long-lasting projects for your home and keep the Japanese joinery and traditional woodworking craft alive!

Intermediate Guide to Japanese Joinery

Introduction

Reading this and implementing its techniques will make you a pro in Japanese joinery. This guide is based on the renowned aspects of traditional Japanese joinery. s. The tools used in Japanese joinery are most important to the beauty of the carpentry. The product of seemingly mystical differences, such as the tool used on the pull-stroke, cannot be undervalued. The wide use of hand planes, and the tempering achievable by their wood frames, is a crucial aspect of how these specific tools stimulate a stealthy craft culture.

This particular aspect gave rise to the closer relationship between material and maker. The extensive use of fire and water as instruments shines effervescent light on the deep knowledge Japanese joinery artists have for their medium. How wood expands when moist is a property embraced by the carpenter.

The topics explored in this book include: Japanese joinery, tools and tool maintenance workspaces, advanced joints, and intermediate skill level shading techniques. When finished, you will have crafted five extremely beautiful projects with precise accuracy as absolutely remarkable decor to compliment your house, office, or any space you desire to embellish. The projects set forth in this book follow progressively, so your skills can continue to advance as they build upon the next. By the end of this book, you will be well-equipped with the fundamentals of Japanese joinery, a further understanding of timber, advanced joint making techniques, and an understanding of the self-discipline that is so essential in the craft. The foundation provided in these teachings is there, so you can take the knowledge and further yourself.

The projects in this book offer details of Japanese joinery and how to craft it. It is a hands-on, step-by-step directive of how these structures are done in actuality by Japanese carpenters. Sharpen your tools and get ready to work with these techniques for your own personal projects.

The philosophies behind the process that Japanese woodworkers adhere to have profound results on the products they create. The most notable of these philosophical schools of thought is having a quiet or silent environment and lightness of touch while crafting. The noise is kept down by keeping pieces apart from one another, and the material is light in weight, a selection mastered by being attentive in choosing lumber, including bamboo.

The philosophies, strategies, and tools of Japanese joinery, and the respect for the skill, have created a lasting and contemporary style. One difference between Japanese joinery and Western carpentry is the philosophy regarding showing joints. Many Western woodcrafters choose to "exhibit" their skill by showing their joint work. The Japanese woodcrafting philosophy is to hide the joints by using wedged tenons and dovetails. Here, the only joints exposed emphasize a joint's labor in holding weight; for instance, in a chair's design. In this case, the joints would be visible where the seat rails are attached to the front and rear legs. The other reason for hiding the joints is so the beauty of the timber remains the most important. Another interesting reason for hiding joints is if they might protrude slightly due to dampness. The would hide the joints at a 90-degree angle, so the line of intersection bisects the angle. The joints therefore, become less susceptible to moisture and can work smoothly all year around.

Circle the answer to the following multiple choice question: Which is a characteristic of Japanese joinery?

    A. Exposing the joints by using wedged tenons and dovetails.

    B. Hiding the joints at a 60-degree angle, so that the line of intersection bisects the angle and the joints become less susceptible to moisture.

    C. One of the hardest challenges in pursuing Japanese joinery can be having the will to embrace all of the teachings.

    D. There is very little need for practicing Japanese joinery, since people usually get it right the first time.

    E. None of the above.

    F. All of the above.

Chapter One: Japanese Joinery, Tools and Workspace

If you looked up Japanese joinery now, you will find a multitude of videos and essays discussing complex work pieces and joints. But this book is not geared for reading about someone else's work; this intermediate guide will aid you in enjoying touching the wood, smoothing the surface with a hand plane, and putting pieces that you have crafted together with your own two hands. Through step-by-step tutorial projects, you will advance your knowledge and skill of Japanese joinery and learn to craft more complex designs than before as a beginner. Don't panic if you think they look complicated—the steps we will go over will be easy to understand. No matter how complicated it may look, the result of every step, when done as instructed, will make it possible to achieve any challenging goal successfully. You don't necessarily have to be super coordinated, but you do need perseverance. Certain wood crafting techniques try to fix disadvantages forcefully by grinding the timber and glueing it while destroying its natural and unique characteristics, basically resulting in chipboard.

Such disadvantages include burles—or *moku* in Japanese—by grinding the wood finely, and then fixing it with adhesives, destroying the inherent, unique characteristics of the wood. The end result of this is known as MDF, or chipboard. Japanese joinery is about getting to know and working with the different personalities of timber. At its essence, Japanese joinery recognizes every piece of timber as a unique, living element and finds a creative way to bring about the best in each. This is the foundation of the woodworking tradition. For example, homestyle meals that grandmothers have passed down through generations are, of course, better than processed and prefabricated meals. Likewise, chipboard has almost no scent from the beautiful piece of timber it once was. It has lost its inherent personality by bending to human will.

One may ask why Japanese joinery uses less adhesives and fewer nails. It is because metal rusts, first of all, and timber starts to rot when the screws or nails rust. Nails are also not sufficient in harboring the movement of the wood. Wood moves constantly with changes in the environment and the seasons. Japanese joinery is closely analyzing the differences in each piece of timber and connecting them, so when they try to move in different directions, they keep each other still and stable.

The problems with using so much glue is that it removes the ability of the timber to breathe. These days, we have developed adhesives so strong that they can pick up a car. That is exactly the problem, since wood is always moving and wants to keep moving. For instance, you glued two pieces of timber together vertically, then glued the edges in place with a strong glue (i.e. super, gorilla, etc.). You may think you did a great job until the middle sections of the planks started wanting to move. They want to expand when it is humid and contract when it's cold. If these changes occur, and the edges of the planks are restricted in movement... *CRACK!* They will probably split right down the center. Moving parts are important for wood, just like they are for people. Most of all, it is very enjoyable to construct a piece through your own ingenuity. Creating a framework formation using a saw, plane, chisel, your imagination, and your own two hands into something that is joined together with other pieces to form one solid structure is a rather satisfying accomplishment.

Modern day engineering of timber is an example of society's will to eradicate anything inconsistent or unpredictable. In Japanese joinery, wood crafting is tightly managed. Most historical buildings in Japan exhibit standards in their construction by repeating glue the elements, and there, century-old structures still remain in good condition. Craftsmen practicing joinery techniques exert mastery over the timber and other such materials' behaviors. However, the way they do it is considerably different from Western industrial methods. Rather than eliminating materials that vary in strength according to the direction of their measurements (i.e. plywood), Japanese joinery crafters go to great lengths to take advantage of these specific traits. Not only does the practice with timber's innate grain by methodically positioning structural pieces make strong connections and prevent sag, but it also utilizes the timber's original state to its advantage, such as using a north-facing tree on the north side of a house (Brownell, 2016). Furthermore, crafters design specifics that project the timber's inescapable transformation over millennia, compensating for deviation and constriction.

Such skill shows profound mastery over timber and extraordinary technological finesse. A common thought is that old structures are physically inferior to modern steel and concrete fabrications, yet this is debatable. First of all, the quality of the materials used are often inferior. A manufactured wood beam pales in comparison to a single log of hickory or cypress. Historical methods for processing timber are more rigorous than modern manufacturing methods. What should be challenged today is the possibility that expediency in contemporary construction contributes to material decline. With the recent rallying of interest in structures made of wood, Japanese joinery offers inspiration for future, wood-based construction methods. Western carpentry would benefit from a better appreciation of timber origins, its deployment, and behavior over long periods of time (Brownell, 2016). Various factors give rise to irregular and complicated patterns in timber.

Moku (Burls)

There is *tama-moku* (circular burl), *budo-moku* (grape burl), *sasa-moku* (bamboo leaf burl), and *uzura-moku* (quail feather burl).

Moku (burls) are bulb-like growths on timber, usually as a result of an injury to the tree (see the image above). These *moku* make for eye-catching accents, specifically as handles, veneers, inlays, and more. For this reason, *moku* are extremely valuable in the world of Japanese joinery. If you are fortunate enough to stumble across a tree that is down and has burls, you can cut large sections of it, including at least six inches of the trunk material either above or below the *moku*. Being encased by the extra timber protects the *moku* from drying out too quickly. Try to save as much of the *moku's* figure as possible. Do not slice it from the trunk—unless it is too big to handle—but slice through the meat and take off the back half of the trunk, which will make it lighter and smaller. For transporting noticeably large *moku*, cut them into manageable pieces while still allowing extra material.

If you are cutting *moku* into planks, cut 25% thicker than necessary, so there is room for change during drying. Also, depending on the grain of the *moku*, the direction you cut can make a change in the appearance. *Moku* usually have a random swirling or eye figure grain. To dry them faster, trim and size the *moku*. If you are not in a rush to use it, you can also slowly let them air dry. With this method, they can produce changes in their color and spalting (change in wood color caused by fungi), adding extra character to the piece. The only way to tell if the *moku* has an eye figure is to trim a thin slick from the top of the *moku*. Eye figures in the *moku* are generally found in walnut, maple, ash, and cherry trees. Swirling grains in moku are usually found in birch, gum, and mulberry. Using end grain sealer for wood, cover the exposed ends. Make sure to store it away from direct sunlight and rain. Thick pieces of *moku* can take up to six months to dry, whereas, ¾-inch planks take four months when stored in a warm environment (Hoover, 2020).

Steps for Cutting Rounds in Moku

1. Using a chainsaw, cut through the pith of the oversized section and remove the back half of the piece, while also making the back of the *moku* flat.

2. Trim any excess trunk ends. Chainsaw the sides of the *moku* to make it square.

3. Make a disc out of cardboard as a guide for cutting the *moku's* top.

4. Using your band saw and cardboard disc, complete the round.

Timber

Wood needs to dry for several years after being cut down. It is then processed (*kidori*) according to its need. In

Japanese joinery, timber can be cut across (*itame-dori*) or with the grain (*masame-dori*). The growth rings in with the grain (*masame-dori*) cuts appear on the surface as a series of parallel lines. When cut across the grain (*itame-dori*), it appears as irregular patterns. When the timber is removed from the outer part of the tree, the grain is glossier and more attractive than if it was taken closer to the center.

Japanese Joinery Tools

Joinery tools are used mainly on a pull-stroke control, rather than the Western push-stroke technique. With a wooden bodied hammer (*Kanna*), adjusted planes are pulled toward the worker in order to make the cut. Likewise, Japanese saws' (*nokogiri*) teeth are angled toward the worker, so they meet with the timber fibers with the pull stroke. When it comes to saws, the pull stroke has an extensive effect on the tool's build. Western saws have to resist the buckling intensity of cutting on the push stroke. The motion of pulling a Japanese saw through the cut results in the blade tensioning naturally. This saw also has better accuracy because of its thin blade. The fine blade promotes precise and delicate cuts. The quickness matched with the precision of the cut advances a wider use of hand-saws used in joinery.

Circle the right answer: Which of the following saws makes cross cuts and bevels?

A. Radial.

B. Compound miter.

C. Hack.

D. Air.

Japanese hammer ("Genno" or "Gennoh" 玄能): There are various types of Japanese hammers; some are used to position hand plane blades, some are used for chiseling, some for pounding out steel that had been hardened by lacquer, and some just for hammering nails.

Dead-blow mallet: A hammer hollowed out and filled with steel. It aids the crafter in delivering controlled and solid blows to the piece. The large head widens the force over a wide area, making it perfect for joints.

Japanese gimlet (kiri 錐): The kiri is used for hollowing out circular holes in a piece of wood, which is usually the first step in boring out a mortise. Although it may seem simple to use, the Japanese gimlet is considered a very difficult tool to master.

Japanese saws (nokogiri 鋸): The *nokogiri* uses a pull stroke to cut, allowing the blades to be much thinner than what Western saws have. The two principal types of cutting teeth are the rip (*tatebiki*) and the crosscut (*yokobiki*). There are various other types of Japanese saws used in joinery too, such as the stiff-backed saw (*jdouzuki*), ridge saw (*azebiki*), press-cut saw, (*osae-biki*) among others.

Inkpot (sumitsubo 墨壺): The inkpot is used for marking lengthy straight lines onto varying wood surfaces. A piece of thread is fastened to a round wood piece, with a needle attached at the end. The opposite end of the thread is inserted through the small opening at the end of the inkpot through a pit containing ink. It is then wrapped around a spool. The ink is kept in an ink-soaked silk wadding (Butler, 2004). The silk threads are used to draw lines by holding the inkpot (*sumitsubo*) in one's left hand, positioning the thread onto the surface, and slowly moving away until the measured degree of thread is unraveled, at which point you would press your thumb down to stop the reeling. Then, with one's index finger, they would press the thread down to the measured end point of the line. Using one's right hand, they would pull the thread upward until it is tense, then suddenly release, leaving the line straight on any surface, even if it has irregularities.

Blades

In Japanese joinery, blades are made by steel smiths and are unmounted, so the carpenter can fashion the handles by their own hands. In Japan, steel has always been crafted with a high level of refinement, for without, the detail and beautiful surface work in Japanese joinery could not be accomplished. Similar to the Japanese chisel and plane, the edge of these blades are extremely thin, hard, and welded to a softer metal piece to function as a shock absorber and protect the tip from breakage. This traditional albeit advanced technology permits the use of harder steel and sharpening of a much finer edge. In Japanese joinery, it is customary to go through at least three whetstones that vary in coarseness, moving from the roughest to the finest stone. The blade of both chisels and planes are noted by the hollow sole on their flat sides. This portion makes sure there is a high degree of flatness when sharpened. Note that when the flat side is burnished, it cannot develop curves or rock, as it is the only contact with the whetstone. This advances the precision of the cuts made by the chisel and ensures a smoother contact with the wedge. It thus produces an even degree of support to scroll the entire width of the blade. The hollow sole also vastly lowers the amount of steel needed to be taken out to acquire flatness on the back side of the blade, lessening the need for resharpening.

When it comes to chisels, the blades reduce the resistance to friction as it is pounded into or taken out of the timber. Also, the interaction between the blade and leading edge is a relationship that changes as the blade is resharpened. With the blades on planes, as the edge is honed down to the hollow's rim, it can then be pushed out—a process called *ura-dashi*, where a pointed hammer pushes in the blade slightly downward along the bevel. The back of the blade is flattened again, after which the hollow is reestablished. This process acts as a gauging mechanism for sharpening and prolonging the life of a thin blade. It also keeps consistent measurements of the blade over time. There are various steel types used for Japanese chisels and planes. White steel (*shiro-gane*) is almost pure steel that resharpens easily. Blue steel (*ao-gane*) is a bit harder to sharpen than white steel, but it is sturdier.

Planes

It is the act of pulling the plane into the body that makes this tool successful. In traditional Japanese joinery, woodworkers would work on the floor, which allowed the worker to generate more power from their sitting position. Sitting also increased the worker's control and enactment with the timber and fosters a more controlled, careful engagement with it. The act of pushing a tool through a material requires an explosive thrust of force. The further away they are from the worker's body, the less control. It is also more difficult to see the shaving made, which is a valuable measure of the cut-quality achieved. Japanese planes are drawn toward the body; therefore, they can control their cutting better, and the shavings are always visible. The plane's wooden body gives the joiner unspoken knowledge. For instance, the density and difficult grains can be felt more through the body.

Knives

Surprisingly, the knivery used in Japanese joinery also employs a pull-stroke. In basket-making, the known technique in preparing the bamboo is for the knife to be gripped against the crafter's knee, and then the bamboo would be pulled along the blade. The bamboo is supported by a block of wood, which helps ensure the thickness of the cut. Pulling the knife makes a thinner and finer medium to be worked with. It also keeps the eye of the user on the exposed blade. Lastly, the maker has a better feel of the stress the knife is subject to and can change their technique accordingly. The pull-stroke is one of the main differences between Japanese and Western woodworking. It is comprehensive in its effect—that is, in the tools' anatomy, how they are experienced, the size and species of the wood, and in what clues are fed to the user during the project. Research on the Japanese pull-stroke found that users felt more control, had better accuracy, required less effort, and were given more subtle information (Ryokogyo, 2020; Sennet, 2008; Toshio, 2006).

Vises

Vises used in Japanese joinery are wedges of timber tied to a pylon with a loop of rope. The timber is then put under the wedge, and the wedge is pounded down with a hammer. Vises are used much less in Japanese joinery than in Western carpentry.

Techniques with Water

Water is actually a tool used frequently in Japanese joinery, unlike in the West, where it is feared. Wood is composed of a bundle of tube-like patterns running up and down the tree. These tubes have a lot of water in them, which adds to its weight. In order to use the tree's gift of wood, the water needs to be removed through aeration or dried in a kiln. This way, the timber is stable and strong. Interestingly, the tubes do not lose their ability to re-take up moisture. When left in a wet environment, dry wood will absorb water until it is equal to the moisture of the environment within which it dwells. That is why crafters usually try to keep their wood from being exposed to moisture. In Japanese joinery, water plays a more harmonized role, evident in the utilization of waterstones to sharpen items, where the moisture lubricates the sharp particles of finely grained sedimentary stone. Today, waterstones are used widely in Western shops and made synthetically in circular forms for use in electrical sharpening equipment. In Japanese joinery, water is used in a much more detailed process other than just sharpening.

Stream-bending

Stream-bending is a technique whereby a piece of wood—if you were making a bowl—would be oval shaped. It is then soaked for three days in water. Then, the wood is microwaved for three minutes, which allows steam to form. This process makes the wood flexible and easily bendable. The timber is then placed in a shallow curve holder to form the shape. When the wood dries, the shape is set, resulting in an elegant serving dish. It may sound easy, but not necessarily. The technique is based on a tried and true method forged in experience and shared throughout the trade (Sugawara, 2020). The soaking time, temperature of the water, time in the microwave, and thickness of the wood are all finely balanced.

A more common use of water is when planing. The crafter always wants to plane in the direction of the timber grain, which helps them avoid tear-out and the necessity of having to use a chip breaker or sharper blade, or needing to change the direction to lower the cut's angle. One method used in Japanese joinery is to mist the surface of the wood and let it sit for several minutes. This technique allows the wood fibers to plump, lending itself to an easier cut. One problem with this method for Western woodworkers is the steel's potential to rust, which is used in manufacturing Western planes. Japanese planes are made of wood, and therefore do not have this problem. Western woodworkers may choose to dust their planes with oiled leather.

Fire Techniques

Heat and fire are interesting techniques used in Japanese joinery. The surface of wood when burned has unique properties. When charred, timber has protection from moisture. Also, the char adds a deep, rich, and dark color. One use of charred wood in Japanese woodcraftery is making use of "charred bamboo." Traditionally, the wood was acquired from the fireplace ceiling in Japanese kitchens. The smoke created by the fire would harden and dry the bamboo roof structures, making it a prized material often seen in crafted chopsticks. The reason smoked bamboo is good for chopsticks is that the wood has become stiff; however, that also makes it a bit difficult to work with.

Tool Modification

It is not an uncommon occurrence to see a one-of-a-kind tool in the kit of a Japanese woodworker. Tools are often personalized and modified to fit the work required and produced. Sometimes, it is just to make them a more comfortable fit. On some occasions, hammer heads are purchased by themselves because the crafter wants to make their own wooden handle. That way, the handle will fit the woodworker's grip perfectly. The woodcrafter may also want to change the distance between the handle and hammer head; therefore, changing the power can then be registered. Plane modifications are most apparent due to the timber bodies that hold the blade in place. Concave furrows allow the plane to move over roughly-sawn surfaces better, thus allowing it to be flattened faster.

Tools used to be very high priced back in the 50's, so woodcrafters had to learn how to do their own metal work. This legacy is still passed on to younger woodworkers through apprenticeships in the present day (Keller, 2020). One of the best features of Japanese joinery tools is that they are cordless!

Sandpaper

It is much more common in Japanese joinery to use blades and planes to smooth and level wood to its final finish. However, sandpaper made from a horsetail plant creates a very fine sandpaper that also polishes. The result of slicing the wood to result in the finished surface is better than a scratched up finish. Sandpaper is quicker, but physically passing a blade over the surface makes their silky smooth finish, which is easily recognizable when compared to sandpaper finishes.

Profiled planes add texture to a wood finish. Woodcrafters and others viewing planned finishes generally have an appreciation for the skill involved in creating it. In essence, it is the crafter's fingerprint. Japanese woodcrafters use sandpaper to remove sharp edges from their finished works.

Tool Maintenance

Ingrained in Japanese joinery is the concept of maintenance, which is, by itself, an art. Of great importance in Japanese woodcrafting is the workers' acceptance of continuous maintenance. This also ensures the passing on of crafting skills to future generations. Timber, due to its ability to become damaged by moisture, has a unique aspect when it comes to maintenance.

Lightness

Lightness in the craft is a theme present in Japanese joinery's philosophical background. Sometimes, this was considered in the species of wood, though other times it was the details employed in the construction or the careful design, which gained the desired effect. Lightness is important to the Japanese woodcrafter because it shows great respect for the value of timber while making it more affordable and efficient. Your craftsmanship should add to the harmony of your home rather than sticking out like a sore thumb. The crafter's search for lightness, in this concept, is the search for harmony that complements a space.

Circle true or false: You can only use hand tools for Japanese joinery.

Recommended Tools

With the right set of tools, you can realistically achieve the skills for the projects described in this book. Practicing joinery by hand will definitely advance your potential and give rise to more confidence with machinery. Although you may already have a nice set of tools the following are recommended:

a. Both a big and regular-sized hammer.

b. Carpenter's square and try square.

c. Bevel gauge.

d. Razor saw *(ryoba nokogiri)*.

e. 24mm paring chisel *(tsuki nomi)*.

f. 24mm striking chisel *(tataki nomi)*.

g. 12mm striking chisel *(tataki nomi)*.

h. 48mm hand plane *(kanna)*.

i. 70mm hand plane *(kanna)*.

j. 36mm shoulder plane *(kiwa kanna)*.

k. Left and right metal plates.

l. Diamond plate.

m. Sharpening stone.

Chapter Two: Advanced Joints

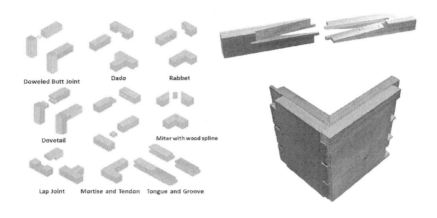

When contemplating wooden construction, it is usually along the lines of horizontal and post-architecture, with all the elements coming together at right angles. However, there are many structures in which the significance is placed on curved surfaces or diagonal elements.

In Japanese joinery, most of the construction uses straight lines. This is primarily due to the nature and characteristics of timber and its growth. The obvious deviations, of course, include curve-shaped kitchen utensils and smoothly curved furnishings, sometimes formed from a single timber piece. In these cases, Japanese joinery is the development of techniques to pressure wood into taking and retaining unnatural shapes. In addition to the vast variety of wood species available, you can pick the timber best suited for your intended use. In this craft, a joint is either a connecting joint (*shiguchi*) or splicing joint (*tsugite*), where in the West, joints are categorized in how the timber pieces are joined. In Japanese joinery, joints are categorized by the way they function. For example, two pieces of timber joined together end to end is referred to as a lap joint, or *tsugite*. A *tsugite* used to join two elements at a 90-degree angle is a connecting joint, or *shiguchi*. A *tsugite* and a *shiguchi* serve more of a purpose than holding wood together; they contribute both to the timber's elegance and its strength.

As you know from the Beginner's Guide to Japanese Joinery, it can be very difficult to get miters to close, or a three-way miter with interlocking tenons you can't even see to close. However, it is inspiring to see a 16-inch long miter connected together with dovetails disappeared, leaving no gaps on the outside or inside corners. Your road toward intermediate competency—leading to mastery—is long, though worth the work. As you move forward to an intermediate skill level, understanding joining systems is necessary to undertake the projects detailed in the following chapters.

Circle the hand tool used to shape timber using only your muscle strength.

A. Bench plane.

B. Bench saw.

C. Electric saw.

D. Saw plane.

Miter Joints

From an artistic viewpoint, the mither joint is an eye-pleasing concept for dealing with corners. It gives the onlooker the sense that there is continuous grain around the corners of an object. It is a sophisticated technique that needs a good deal of precision to be well-executed. Most frames, moldings, and other structures meet at a right angle of 90 degrees; therefore, 45 degrees would be the most common agle. In order for them to look good and be effective, they have to fit together perfectly. The most minor inaccuracy will not only affect the function but the strength and appearance of the miter joint.

Constructing a miter joint is something a woodcrafter of intermediate skill should no longer be struggling with. Any structure made using two or more pieces of lumber has a miter joint. It is constructed by joining two edges at a 45-degree angle on the face of the timber. This way, the timber's end grain stays hidden. Exactness is crucial, and a perfect miter calls for a careful setup with almost monotonous adjustments to follow. Most interestingly, with Japanese joinery techniques, a hand saw will be more accurate because cutting miters with machines causes forces on the material to push it away from the saw blade, which then causes the piece to move slightly. If this happens, it will affect the precision the crafter is working so hard to attain. When hand-marking your lumber for a 45-degree cut, use a miter box and tenon saw to cut the angle. Then, use a miter shooting board (at one edge is a shallow, wide rabbet where a plane rides; the rabbet bed supports the side wing, and the shoulder guides the plane) and a bench plane to rig the face of the miter, then adjust for inaccuracies. The many tools used for constructing miter joints will depend on your project. For a basic cut, a miter box and hand saw will be all you need. For more complicated miter cuts, various types of saws will come in handy, such as compound miter, radial-arm, table, or circular saws.

Tip: While a structure's mitered corners may look great, they can be difficult to clamp. A remedy for this situation is to add a spline across the joint. A spline will keep the joints aligned exactly during assembly.

Pinned Right Angle Miter Joint

Two conditions must be met for cutting clean and tight miters. At each end of the joint, the angle cut has to be the same and add up to the corner angle desired (a 90-degree angle needs two 45-degree miter cuts) and, for a rectangular or square project, the opposing pairs of parts' length must be exactly the same. You may find yourself with troubling, out-of-square corners (i.e. bumps on the surface). The trick to creating tight-fitting miters is understanding how to adjust your cuttery to real-world situations.

Miters usually need to be shaved in order to fit. One method is to adjust the angle just a bit on your miter saw, or simply place a thin piece of material against the saw fence to readjust the angle. Move the material closer to the blade for large adjustments and closer for smaller adjustments. Remember to cut the pieces exactly the same. Regardless of the technique or material used, accuracy is of utmost importance. Discrepancies—no matter how small—in the angle of mitered pieces can end up causing gaps in the joints, either at the base or in the tip. To avoid mistakes, cut a pair of samples. Put the sample pieces together with the biggest angle gauge you have and check your cuts. If they are off at all, there will be a visible gap in the corner.

Mitered pieces can be fastened together with glue, as both the plywood and solid pieces have partial saw blade tooth markings. Using glue actually produces enough strength to hold together items such as picture frames and small gifts. However, using glue can be tricky, since pieces can slip from their positions during the process. Using clamps to apply pressure to all sides is one remedy. When using pipe or bar clamps, it is helpful to drive a small nail into the joint at each corner to keep the parts in place.

Biscuit Joints

One of the lesser-known joinery joints is the biscuit joint, most likely because there are fewer places where you will find this joint to be more useful than other types. The joint got its name due to how the pieces of wood resemble the shape of a biscuit.

These particular joints strengthen and prevent the joint from moving. The biscuit joiner is a tool used to make the round (biscuit) shaped holes in the timber. The biscuit joint's main use is to connect wood boards together, such as table tops, and once secured, they are rather pleasing to the eye. They are quick to cut and easy to mark, but they may not be as sturdy as some of the other joints discussed in this book. Despite that, they are still suitable for many projects. They are strong, versatile, and used in various types of assemblies and for alignment problems. The best way to appreciate this versatile joint is by putting it to use. The size and shape of the biscuit joint makes it better than a dowel. It has an oversized slot, which allows the biscuit to move slightly, allowing you to adjust parts into their perfect alignments. However, once they are glued, the joint will swell, and the parts will be locked in place. The safest way to cut biscuit slots is with a biscuit joiner. This tool is designed to do nothing but plunge-cut arc-shaped slots. Unlike the router/biscuit-cutting bit combo, the biscuit joiner's cutter retracts inside the tool as you pull it away from your work. From there, a flat, football-shaped piece of compressed wood—called a biscuit—would fit into a pair of matching slots.

How to Structure a Biscuit Joint

1. Line the two wood pieces you are planning to join together up carefully on your bench. Use a pencil to mark where you will place the joints.

2. Once you have made the pencil marks, secure the wood to your bench or in a vice, so it is held steady for the cut. Manage your biscuit joiner for the size of the biscuits you will need for your project.

3. Cut into the wood on the marked lines.

4. Make sure your planks join together neatly and tight, then insert the wood biscuits and, while clamping the wood, bond the planks together. Note that the wood biscuits can be bought in hardware stores.

Biscuits

Three-way Joint (kawai tsugite)

Most joints are created with only one or two directions in mind. The *kawai tsugite* joint fits together perfectly in three ways. Assembling the *kawai tsugite* joint either goes as two corner joints or one straight one. At the intermediate skill level, it is important to have a grip on the principles detailed here, as it is one thing to understand how this joint works, but another one completely to cut it into a piece of timber. It is difficult enough to cut a rectangle or square into this new shape, let alone doing it well. The end result of crafting this joint is nice to look at, but due to how the joint reacts with the woodgrain, it is weaker than other types of joints detailed here. However, being at the intermediate skill level, one must know how to craft the *kawai tsugite*. There are different ways to assemble this joint; as mentioned earlier, either two corner joints or a straight joint. First, the tools you will need include:

- A perfect square of timber.

- A back saw.

- Chisel (skewed or straight).

- Marking gauge.

- Bevel gauge.

- Try-square.

- Miter-square.

- A pencil or marking knife.

- Vise

- Bench hook.

- A workbench.

Steps

1. Plane your square of wood to be entirely round. They have to be the exact same dimensions in thickness and in width. You can also cut one piece in half; that way, the grain will be the same for when they are put together in a straight line). Mark a perfect cubic square with a colored pencil (blue). Then, mark 45-degree lines from one corner to the other corner (red). Finally, mark the lines from the midpoint to the corner (green).

2. Along the 45-degree marking (from corner to corner), cut off the corner of the post.

3. Mark new lines on the cut face by connecting the points on each.

4. Using a saw to cut as much of the timber as possible. This will give a smooth face for chiseling. It will also save you a great deal of time.

5. After the cut, chisel out the remaining wood between the cut portions.

6. Put the pieces together. Note that the first try, you may find that you have to chisel a bit more to make them flush.

Dovetail Joint

At the intermediate skill level, you must become proficient in dovetailing, as it is an effective technique when it comes to joining boards together.

Dovetail joints, although beautifully crafted, are also somewhat complicated and created in the shape of a bird tail for a reason. These joints are known for their tensile strength, meaning that they are resistant to being pulled apart. Working dovetail joints are a sign that you have acquired true, intermediate Japenese joinery skills. Because they are free of mechanical joints, they are also much more pleasing to the eye. Important evidence to back up the stealthy nature of the dovetail joint can be observed in almost any antique shop. One need only open one of the drawers on an old desk to see a sure sign of their quality. Dovetail joints are the strongest joints in all of Japanese joinery and will hold together with or without glue. This is the perfect joint for any intermediate joiner to master, as the challenge in creating them lies in their difficulty to mark out and cut. If cut incorrectly, they will lose their advantages previously described. Making sure you design the right type of dovetail for your projects is crucial in building strength in the creation and paying noteage to your skill set and abilities.

The most common usage of this joint is in where the sides dresser drawers are joined. A series of "tails" cut from the end of one plank interlock with a series of "pins" that have been cut from another end of the plank. Together, they form a trapezoidal shape. Dovetail joints have been used in Japanese joinery for thousands of years as a method for joining the front and sides of an item. There are fewer nails, fastenings, and rivets needed because the corners of the joinery can be secured with the dovetail technique. They create cohesion by interlocking the specific cuts, pins, and tails needed to form it. It might be smart to practice with extra wood to see which technique suits you the best. That way, you can try different types of dovetail joints and avoid mistakes in the bigger project. Using the old-fashioned way to cut dovetails gives a unique, handcrafted appearance to your project. With patience and lots of practice, you can master the fulfilling skill of hand-cut dovetails.

The one disadvantage the dovetail has is in comparing it to a box joint. The dovetail joint has fewer tenons than a box joint and they do not go all the way through a piece of timber. In addition, dovetails are used for pulling away from a piece in one direction, whereas a box joint is strong in both directions.

Tools Needed for Dovetail Joints

- Chisel set.

- Marking knife.

- Dovetailing saw.

- Mallet.
- Square.
- Vise.

Cutting Dovetail Joints

With just a little practice, cutting dovetail joints can be fairly easy. Make sure your tools are maintained and ready; it doesn't matter how good your technique is if your tools are not ready for service. Secure your timber with your vice and mark how deep you will cut. You can use your marking gauge for this.

Mark your tails, from the end of the board to the top of the square and the face of the board, and the areas you will be removing. From where you've marked, cut out the tails. If need be, you can mark the cutting angle until you have a good feel with your saw. The following are steps for clearing out spare wood:

1. The best method for clearing out spare wood is by chiseling approximately 1/16th of an inch, as close as you can to the depth line.

2. Make a quick downstroke along the depth line. From there, make a 30-degree angle pair back to stop the most recent cut.

3. Continue this process until you're halfway, then flip your piece over. Keep cutting until you break through.

4. Chisel into the stop line to remove any extra wood. Continue all the way across the line.

5. Repeat this process at the halfway point, then flip your board over. Continue the process once more until you break through to the other side.

6. Mark the pins. Like with the tails, use a single board to cut the pins necessary for the formation of the dovetail joint.

7. Place the pin piece in your vise with the end sticking out about an 1/8th of an inch. Hold a block aligned to the projection of the face. Using a tail board, place the end of the tail piece in tight to the block and align the edges. While applying pressure, mark to the end of the pin board, like you did with the tail board.

8. Cut out the pins. Like the tails, your square will come in handy. Mark vertical lines for your cuts. Then, clean out any waste wood.

9. With your vise, rotate the pin board 90 degrees before you cut to the depth line. This will loosen debris from around the edge of the face. Clean any waste by chiseling, just the same as you did with the tail waste.

10. Fit the boards together (you may need to make some final chisels). Gently chisel any excess wood until you have a perfect fit. Make sure there are no gaps.

11. Now the boards, pins, and tails are set for a perfect fit. Apply some force and lock the boards, forming a corner.

Circle true or false: Intermediate skills include using dovetails of intersecting corners.

Finally, it is time to clean up. It is imperative that you clean off any waste, so it won't ruin your dovetail. A clean workplace ensures a smooth process from start to finish. Having less clutter around will also help you focus more. With this important step of cleaning, you are well on your way to successfully producing the type of dovetail corners you desire. Dovetails are strong joints that will stand the test of time.

Types of Dovetails

A. *Through dovetails*: The simplest and oldest type of dovetail used in joinery. When cut, two separate pieces of timber intersect, making a perfect corner that fits securely and firmly.

B. *Sliding dovetail joints*: Work with a long tail with carved grooves. These dovetails' sides are also angled, so it can adhere to a longboard pin that keeps it in place. Sliding dovetail joints are stronger than dado joints because they do not depend completely on glue. Also, the dovetail shoulders keep the edges of the slot hidden and pay tribute to the visible art of good craftsmanship.

C. *Box-joint dovetails*: Use simple straight cuts for pins and tails, but can be less durable.

Box Dovetails

1. The difference between a *dovetail* and a *box dovetail* is that the digits on the box dovetail are rectangular

rather than shaped like a dove's tail. There are different ways to cut box-joint dovetails; for example, you can do it with a chisel and a dovetail saw by deciding on a width for the fingers, which would divide uniformly into the width of the wood for the project. So, if your wood's width is 6 inches, a ½-inch wide finger would make 12 in total, 6 for each piece of wood. You can also cut them by machine if you do not want to take the time to cut them by hand. Almost all dovetail jigs cut box joints. Look at the documentation that came with your jig—it should describe the process as the same as cutting the tails of a regular dovetail joint, but using a straight cutting bit instead of a dovetail bit. An even easier way is to use your table saw with an arranged dado set. The cut depth should be the same height as the wood's thickness.

2. Adhere a piece of scrap wood to your miter gauge. Make sure it is wide enough so that when connected to the miter gauge, it extends past the blade by an inch, and 2 inches past the miter gauge on the left side.

3. Make sure the scrap wood is square to the blade, then pass the scrap wood through the saw.

4. Move the piece of scrap wood to the right by twice the width of the fingers after removing it from the gauge, then reattach it to the miter gauge.

5. Cut one small piece of wood the same width as the fingers, which will connect into the cut made in the scrap. Note that this piece should be two times as wide as the piece you are working with.

6. Adhere this piece into the notch you made in the scrap using a wood screw from the bottom, and make it so that it sticks forward from the scrap toward the blade of the saw. This will be your finger cutting gauge.

7. Make sure the miter is square to the blade and make a new notch in the current position on the scrap wood.

8. Now you have made the jig. Take one of the pieces you are working with and place it firmly against the scrap on the end, up to the small guide you attached to the notch in your scrap piece.

9. Hole the wood firmly against the scrap piece and guide it through the saw. Once clear, run the whole assembly back through the blade. Now you have your first notch and finger at precisely the right width. Continue this process until all fingers for your project have been formed.

10. The opposing piece of wood is cut much in the same way, except the initial cut is not made with the wood against the guide. The outside edge of the opposing piece should be even with the dado blade's outside edge. This is easily positioned by aligning the outer edge of the wood piece with the edge of the notch in the scrap piece. Once you have lined it up, make an initial cut, continuing until complete.

11. Once both sides of the joint's fingers are formed, dry fit the joint.

12. Next, put a thin layer of flue on the surface of all of the joints, then guide them together and clam if necessary. This technique is fine for crafting drawers

or making boxes. Note that you should be diligent when it comes to clamping the box square.

The box joint is a relatively strong and useful joint that can be a lot of fun to build. It is not as elegant as dovetails, but certainly very appropriate in some circumstances.

Half-blind Dovetails

This cut realizes only one visible dovetail joint, which is not the best for large furniture. It is a joint that does a superb job in its strength and beauty without begging for attention; this is the nature of the half-blind dovetail on a well-made dresser drawer. This is for the intermediate crafter and one who has mastered all other dovetails. The half-blind dovetail is cut the same way, with the difference being that half-blind pins need careful sawing and additional chiseling. The following are steps for a dresser drawer, which you will be making as one of the projects in this book.

1. Cut the groove for the bottom of the drawer in the front and sides. Note that once the joint is assembled, you will be unable to see the ends of the tails. All the groves in the sides of the drawers can be sawed all the way through.

2. Next, you have to decide how long the tails will be (how near they come to the drawer front's exterior). For drawers that are ¾ inches thick on the front, make the tails approximately ⅝ inches long, but leave a ⅛-inch lap. If you make the lap of the joint much thicker, it will look awkward. If you make it too thin, you might chisel through it when you are cutting the sockets for the tails. It is a 5 to 1 ratio.

3. Put the marking gauge edge up against the inside face of the drawer front and adjust your cutter by leaving a ⅛-inch lap. Place a mark on the drawer front's ends. Then, run the gauge's fence—leaving it at the same setting—along the edge front of the side of each drawer, marking the faces on the inside and outside.

4. Reset the gauge to the thickness of the sides and draw a line on the drawer at each end (front and inside). Set the gauge with a cutter just overhanging the side, so the pins will be a bit longer than the side is thick. That way, they will be easier to sand or plane flush after assembly.

5. Lay out the tails on the side of the drawers and cut them the same way as you do for through-dovetail joints.

6. Reset the gauge to the thickness of the sides and scribe a line on each end of the drawer front, on the inside face. When I set the gauge for this line, I let the cutter overhang the side; that way, the pins will be slightly longer than the side is thick, and they'll be easy to plane or sand flush after assembly. With the marking gauge work finished, lay out the tails on the drawer sides and cut them, just as you would for through-dovetail joints.

Mitered Dovetail Joint (kakushi arigata kumitsugi)

The mitered dovetail joint is also referred to as the secret miter dovetail joint—the pins and tails on this joint are hidden and when assembled, it appears as a miter joint. This type of joint is used primarily because of its strength, since when it is finished, it will have the appearance of a miter joint but the strength of something bigger. The following are items you'll need and steps to take in order to make one.

1. Two pieces of 6-inch wide pine (it has a straight grain and is not too difficult to work with). Be sure there are no knots where you will be cutting the pins and tails.

2. Set a wheel marking gauge to the thickness of the timber.

3. Use another wheel marking gauge set to 3/16 inches.

4. Lay out your timber so you can match the grain and label the inside and outside.

5. Use a marking gauge set to the thickness of the timber and draw lines on the interior of each piece.

6. Use a marking gauge set at 3/16 inch and draw a line on the interior of each piece.

7. With the same marking gauge, draw a line on each plank at the end, off of the outside (this is where the rabbet will be cut).

8. Mark the line around the sides (outline the rabbet).

9. Here lies the challenge—using a miter saw, set so it won't cut all the way through the plank. To do this, adjust the miter so it leaves 3/16 inch of the rabbet's bottom. The cut has to be neat and accurate.

10. This can also be done with a hand saw or miter gauge and table saw. If the saw's blade doesn't make the bottom flat, don't cut it all the way to the scrib. Clean

the bottom up with your chisel or router plane (see image below).

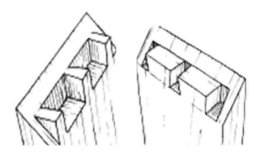

Cross Lap Joint

A lap joint can either be a full lap or half lap joint. With a full lap joint, none of the material is taken from each of the pieces that are to be joined, resulting in a joint that is the thickness of two pieces. With a half lap joint, material is taken out from both pieces, so the end result is the thickness of the thickest piece.

In Japanese joinery, a cross lap joint is when glue joins two long-grain timber faces. It is one of the strongest in its ability to defend shar forces—more than even mortise, tenon, and other well-known "strong" joints. Also referred to as overlapping joints, cross lap joints are used to design continuous, beautiful lines in furniture. Cross lap joints are akin to what you would see in the joinery of log cabins. Each log of timber interlocks neatly to the next without the need for adhesive or nails. However, you want to use nuts and bolts for additional reinforcement.

The edges of cross lap joints are completely level with the surface. Knowing how to achieve cross lap joints in joinery is imperative for making frames or furniture. You will need to take out a section of wood from each adjoining wood piece, and then clamp them together. The difference between a cross lap and half lap is that the joint sits in the middle of either one or both pieces, rather than at the end. The two pieces are at 90-degree angles to each other, and one piece may stop at the joint or carry on past it. When one piece stops at the shin, it is called a middle or "tee" lap. In the cross lap, when both pieces go beyond the joint, both pieces have two shoulders and one cheek. This joint is commonly used for framing cabinets and for bracing items.

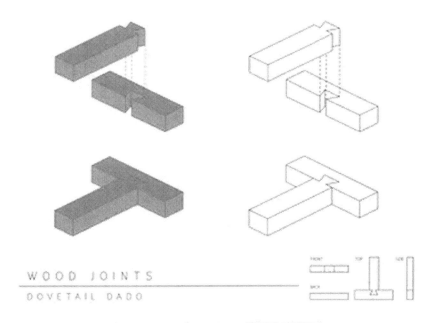

shutterstock.com • 379990726

Tools Needed:

- Table saw.

- Router.

- Jigs.

- Marking gauge.

Steps

1. **Square lines**—With your knife, face mark and square the pieces.

2. **Square**—With the upper parts against the squares, lay the pieces on each other by taking off the square and pencil marking the points at the sides for width.

3. At the marked points, square a line across your timber on both of the pieces of wood making sure to mark halfway across the edges.

4. Set the marking gauge for one-half of the woods' thickness.

5. **Saw**—Chisel a V-shaped groove to start the sawing process and run your saw against the line, making sure the cut is accurate.

6. **Wood waste**—Remove excess wood by making a couple of cuts, and then chisel.

Isuka-Tsugi (crossbill joint)

The crossbill joint (*isuka-tsugi*) is used to secure a horizontal piece of wood to the face of a square pillar. It is called a crossbill joint because it looks like a bird's bill. Make cuts on two corners of the same side of the pillar, leaving the center section uncut. The horizontal beam has a single notch cut so it can be hammered into the flat center part between the two notched corners until secured. The uncut part fits into the gaps in the corner.

Isuzaka Shachi (Crossbill + *Shachi*)

Shachi is a thin peg inserted into a hole where it will receive the corresponding tenon (projecting piece). This technique strengthens the joint. In Japanese joinery, there are two very different types: crossbill joints and scarf joints. The type depends on whether the joint's face interlocks or not. An ordinary scarf joint is just two flat pieces connecting on an angle relative to the axis of the piece being joined. It depends wholly on the glue and/or screws, bolts, and nails. Interlocking crossbill joints vary depending on the degree of tension and compression strength, but most crossbill joints need mechanical tools to keep them closed. A plain scarf joint is not the joint of choice when strength is needed, so it is used more for decorations. However, with the advent of "super glues," the structural performance of a plain scarf joint has been enhanced.

The joint is made by cutting opposite tapered ends on each piece when they are connected together. For crafters, this makes for a better long-grain to long-grain surface to glue, which makes a stronger joint than one could achieve through crafting a simple butt joint. The tapers are usually cut at an angle ratio of about 1:10. The plain scarf's ends are separated by tiny lines to a fine point, which helps hide the joint of the finished project. By contrast in other scarf forms, the ends are frequently cut to a blunt nib, which connects to a matching shoulder of the partnering piece.

A crossbill joint can also be used to mend problems made when a plank has been cut too short for the project.

The plank can be halved with a tapered cut, then make a scarf joint. When the joint is glued together, the tapers slide against one another making it so the two pieces are not any longer in line with one another. This, in effect, can make a longer board. Once the glue dries, the plank can be sanded down to an even thickness, meaning it's thinner but longer.

Koshikake-Ari-Tsugi (half-lap dovetail joint)

Shaped much like a trapezoid, the half-lap dovetail joint used to couple horizontal pieces. This combines two joints, a half-lap and a dovetail, making an end joint. The dovetail hole and the bench part of the Japanese joint are cut so the hole is approximately half the thickness of the wood. The seat or bench made from the half that remains sticks out like a step beyond the hold.

The second piece of wood has the dovetail projecting piece. Here, the undercut extends over to cover the bench partially when the dovetail projection is placed into the corresponding hold. This makes a face-grain-to-face-grain joint with lots of surface to glue. Butt joints differ, in that they depend on end-grain-to-end-grain connectedness that can break easily. Even if a butt joint is reinforced with a dowel, they are still not as strong as a half-lap dovetail joint.

This provides a face-grain-to-face-grain joint with plenty of gluing surface. Simple butt joints, on the other hand, rely on an end-grain-to-end-grain bond that can break easily. Even a dowel-reinforced butt joint won't prove as strong as a half-lap. Half-laps are commonly used for outdoor furniture, dressers (internal pieces), leg frames, and even your shop-cabinet door frames. You only need a radial-arm or table saw to craft a half-lap joint. For faster and smoother results, you can also use a dado set. However, if you don't have one that will cut cross grain cleanly and leave the surface flat and smooth where you sawed, you should use a router table equipped with a straight bit.

Steps

1. Install your dado to 13/16 inches, so you can achieve the widest possible cut.

2. Calibrate your rip fence, where one edge of your wood piece bumps against the fence and the opposing edge aligns with the dado side calibrated farthest from the fence.

3. You can keep your fence calibrated to this position for the rest of the cuts, provided all of the pieces are the same width. If they are different widths, note that you can use a piece of another adjoining piece to calibrate the fence for the half-lap cut (i.e. horizontal pieces are used for a door frame to calibrate the fence for the cuts in the vertical pieces, and vice versa).

4. Calibrate the cutting depth of the dado unit to where it takes off exactly half of the wood piece thickness. Use some scrap wood to test cutting depth that is of the same thickness as your work pieces. Note that after you cut the scrap wood, you need to place them on a flat surface and line them up so the top and bottom faces are smooth together.

5. Place a mark on the face side of your wood pieces, so as to not confuse them. Note that you will have to place the face side of one piece and the face side of the connecting piece down.

6. Attach an additional wooden fence to your miter gauge, which should be calibrated for a square cut. The added piece should be with a ½ inch of the touching side against the rip fence.

7. Place the wood pieces in position, with an edge up against the added fence and one end pushed against the rip fence.

8. Holding the wood piece firmly against the added fence, pass the piece through the dado set, making consecutive passes until you finish the half-lap cut.

9. Apply wood glue to all of the surfaces to be matted. Bring the wood pieces together with bar or pipe clamps.

10. Put the glued surfaces together tightly with a smaller clamp. To clamp the joint, first apply wood glue to all mating surfaces. Draw the workpieces together with bar or pipe clamps. Then, bring the glued surfaces tightly together with a small clamp.

11. On the faces of the joints, put a piece of scrap wood on, so they are protected from the jaws of the clamp.

12. Every so often, you need to put a half-lap joint on the piece somewhere other than the end. All you would do is mark with a pencil the overlap position onto the piece's edge. The piece will be cut in its center section.

13. Calibrate the edge of the unmarked piece against the additional fence of the miter-gauge. Line up the pencil marks with the dado set sides and put two clamps (handscrew if possible, which is used for fastening a small piece on the drill press) on them. If you do not

have handscrew clamps, just clamp together two wood blocks with bar clamps. If you have them situated correctly, the stops will restrict the area of wood that has been removed from the space between your marks.

14. Align the other end of the wood pieces against the stop that remains and cut another one.

15. Remove all the wood material in between the two cuts. Now that you have your stops set up correctly, your pieces will all turn out to be the same.

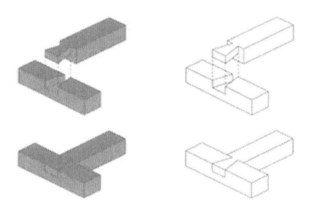

Shiho-Kanawa (all-direction iron ring)

The *shiho-kanawa* is a mortised, rabbeted, oblique scarf joint, able to withstand weight from all directions. This is a notched cut on the end of the board that can support weight from all directions. It should be noted that this all-direction iron ring is a bit strange, in that it is hard to tell which direction it has been inserted from. It is configured by inserting the joint diagonally.

Both ends of the joints are the same and labeled upper and lower wood. Two holes are made deeper through inserting draw pins through the depth of the splice. The joint is crafted by moving the inside face of the upper wood over the inside face of the lower wood while keeping the surfaces of the middle face in close contact. The wood pieces are then secured together by two draw pins, successfully interlocking the back and front surfaces of the joint. The pins are put in from the thicker end toward the thinner one, alternating the process.

Shikake-Ari (locked dovetail joint)

The locked dovetail joint has a projecting tenon partially removed. This area is where the other section is secured with a dovetail joint, which is usually reinforced by combining it with other joints, so it can bear the stress of other loads.

When connecting squared wood pieces, you want the joints to be strong. Strength is acquired by crafting varying squared wood pieces, so if one dissipates, the design can remain strong. The use of ironware commonly used in carpentry today bears the burden of potential corrosion and coming loose as the timber ages. All matter decays with age; however, the decayed area can be removed and replaced with a new section. For this reason, Japanese joinery structures are some of the world's longest-lasting ones too.

Shiho-Henkei (omnidirectionally transformed joint)

Timber is alive and should be used in such a way that helps preserve its strength and natural beauty for as long as possible. This all-direction joint utilizes a differing joint on each side. When it is transformed, it is hard to tell how it has been connected.

Chapter Three:
Project 1—Simple Stool

There is great satisfaction in creating something and completely bypassing the furniture stores. With your Japanese joinery tools in hand, you can be the master of your imagination. You can spend $50 on the materials you need for a piece of furniture that could cost upward of $500 to buy from a store. But remember, as with any applied skill, much of that success comes through trial and error. Wood will not forgive you for any mistakes made, so think carefully about each step and before you start a project. Think of the following saying: "Consider your timber species and ask yourself if pounding a nail into your 2 x 4, will it crack?" Have you practiced sanding enough to learn what affect the varying grits have on different surfaces?

The most essential aspect of Japanese joinery is patience and enjoyment. Focus on that as much or even more than the end result. If you focus on patience and enjoyment, you will go from intermediate to advanced woodcrafter before you know it. With all the projects set forth in this guide, you will be taking measurements. Here is a tip: stick a roll of masking tape in your toolkit or pocket and jot your measurements on the tape. Then, tear it off and stick it where it belongs on your project (stool, chair, table, dresser, shelves, etc.). Something as simple as this can make the experience so much more enjoyable.

Whichever project you are working on, remember that if you measure twice, you will only have to cut once! You will be starting with a simple stool, and as the projects proceed, they will gradually get a bit tougher as you gain more experience in your craft. Let's get started!

Materials:

1. 2 x 2 at 8 feet long (1).

2. 1 x 2 at 10 feet long (1).

3. 1 x 12 at 11½ inches long, or plywood around 12 inches long and ¾ inches thick. (1).

4. 1¼-inch pocket hole screws.

5. Sandpaper.

6. Wood filler.

7. Wood glue.

8. Stain.

What to Cut:

1. (4) 2 x 2 at 23¼ inches long, cut ends at a 5-degree angle and bevel(4) 1 x 2 at 7 inches (short point to short point) cut ends at a 5-degree angle. They are not parallel.

2. (2) 1 x 2 at 10 inches cut same as above. They are not parallel.

3. (2) 1 x 2 at 8¼ inches; same as above.

4. (2) 1 x 2 at 9½ inches; cut same as above.

5. (2) 1 x 2 at 8 inches; cut same as above.

6. (1)1 x 12 at 11½ inches; make the cut to a square by measuring the width of your 1 x 12.

Steps

1. Set your saw to a 5-degree angle and a 5-degree bevel. Trim as close to the end of the 2 x 2 as possible.

2. Measure 23¼ inches and make a cut along the same edge. Now you have one leg.

3. Do the same by sliding the board down as you make more cuts until you have all four cuts out of the 8-foot long 2 x 2.

4. All ends are to be cut at the 5-degree angle for the rungs and supports (no bevel). Different from the legs, the ends will *not* be parallel.

5. Make a 5-degree cut, rotate your board 180 degrees (like you are rolling a rolling pin) and measure your cut from short point to short point.

6. Make another 5-degree cut.

7. Mark the inside corners of the legs, noting that these corners are on the inside (so you don't lose track of the correct leg positions).

8. Cut four top supports at 7 inches long, with both ends at 5-degree angles (not parallel; short point to short point).

9. Attach two legs to the top support.

10. Using pocket holes, build two of these. Center your top supports.

11. Add additional rungs for support.

12. Connect the two leg frames to the remaining top supports.

13. Connect the top by screwing through the top supports to the bottom side of the top, so you hide the screw holes.

14. Use wood filler to fill the holes and let dry.

15. Using 120 grit sandpaper around the top until it is comfortable to sit on.

16. Finally, flip it over and burnish to match your dining room.

Chapter Four:
Project 2—Three-Shelved Rack

Design your own three-shelved rack and base it on your available space. When making your shelves, measure the space available; don't go thinking that all areas are square. You will want to add some vertical sections every 30 inches and give or pulse so your shelving doesn't sag. For a normal, 1-inch wood piece, do not use more than 18 to 24-inch spans.

You can return and cut 1-inch square side panels and place them vertically on the lower shelf for added support if need be. Crafting your own shelves is a wise investment. Using your Japanese joinery skills will return furniture that will appreciate in value with time, due to its high quality. By creating your own shelves, you will have a piece of craftsmanship that will maintain its beauty and value as an inspired work of art constructed by your own hands. Building shelves are one of the most common joinery projects. Building shelves that need to support a substantial amount of weight to stack books or even hold an audio system can be a bit tricky. Using a 1-inch thick piece of plywood will most likely sag in the middle under a considerable amount of weight.

Three-shelved rack

Materials:

1. 1 x 12-inch timber. Have your measurements for the outer box, shelves, and vertical supports. Note that timber comes in 96-inch sections, so you will have some scrap left over.

2. Poplar or maple wood; other species have knots, making them harder to work with.

3. 1 x 2-inch supporting strips, multiplied by the length of your shelves.

4. ⅛-inch plywood for the back.

5. Four supporting pins for each shelf.

6. Wood or drywall screws.

7. Small wire brads for mounting the back.

Steps:

1. Measure the wood cut to fit the shape of the basse box. As a top, you can add or subtract the wood's width, depending on which way your joints lap.

2. Put the box together and place on the plywood backing.

3. Measure every vertical section again, and then cut.

4. Drill the holes necessary for the project.

5. Prior to mounting the vertical box sections, cut dados to mount in the shelves.

6. Take measurements of the pin diameter.

7. Drill the necessary holes. Before mounting the vertical sections in the box, either drill holes for the supporting pins or cut grooves (dados) to mount the shelves into. Using a caliper with a size fitting the diameter of the pins and choose a bit for a loose fit (use scrap for trial and error).

8. Drill perpendicular to the wood holes. As an added tip, you can use a template for hole marking.

9. Stage the vertical sections.

10. Double check the joints to make sure they are square.

11. Measure the length of each shelf and cut.

12. Insert the pins for a test fit.

13. Take out the shelves and add a 1 x 2 inch supporting strip up front for added support.

14. Put the shelves back in.

15. Use 1½ strips of wood along the sides, top, and vertical supports if you want to cover the joint work.

Option 2: Breadboard Edge

- Flatten one side of the board with your plane. First, shave the raised spots, then plan from one end to the other.

- After you have figured out how long you want your shelves to be, the breadboard ends would be about 2-3 inches wide. You can make them wider or narrower if you wish. Then, make the mortise (hole) about ⅔ the width of the end piece.

- Cut the tongue (protruding side) approximately ¼ narrower than the length of the mortise in the end piece. Next, drill the peg holes through the middle of the protruding end piece (tongue) length.

- The length of the ends will equal the width of the wood panel at its widest (to account for humidity). If you are in a cold and dry environment, make the breadboard ends a bit longer than the width of the wood panel.

- Mortise the end sections.

  - Center a ¼-inch mortise (cut a hole) 2⅛ inches deep on one edge of each end piece. Make sure to stop ¾ inches from the end. Drill the holes at both ends first, then drill evenly spaced holes in between each end.

  - Clean the holes out.

  - Create overlapping holes with a drill, beginning at the ends. Afterward, make the sides and ends straight and smooth with a chisel and square the corners.

  - Center a ¼-inch projection strip (tongue) 2 inches long on each end of the board.

  - Using a guide block to keep the saw in line and cut the laid-out shoulders crafting the tongue width at ¼ inches. Lay out and cut the shoulders to make the tongue width a ¼ inch less than the mortise length.

  - Assemble the end pieces, centering on the board width. One inch from the joint, along a line, drill ¼-inch holes ⅝ inches deep in the center and 2¾ inches from each end.

Chapter Five:
Project 3—Cupboard with Two Shelves

Making your own cupboard will give you the ability to personalize your work to what you want. The quality of your cupboard will depend on the quality of your materials. Since you are crafting it yourself, your cupboard will meet your specifications and should have a higher quality of wood. Customize your cupboard to fit your design style, storage inclination, and lifestyle. By deciding on the design that will go into your cupboard yourself, it becomes an even better representation of you.

Step 1: *(see diagram below)*

The body of this cupboard is crafted with ¾-inch plywood.

A. Cut each long side.

B. Cut the bottom and middle shelves.

C. Before attaching, add your shelf pin holes to the sides.

D. Adhere the shelf boards to the sides with mortise and tenon.

Step 2:

A. Adhere the back support plank to the top and the back of the cupboard. This plank attaches the cupboard to the wall when finished.

B. Make sure to mark which side of the line you will set your mortise.

C. Adhere this plank to the side boards' mortise and tenon.

Step 3:

    A. Build the frame face.

    B. The top and side planks are 1 x 3, and the bottom board is 1 x 4.

    C. Mortise holes on the bottoms and tenons on the tops of the side boards.

    D. Connect all the planks to construct the frame.

Step 4:

    A. Connect the frame face to the front of the cupboard.

    B. Place your mortise and tenon on the inside of the cupboard and through the frame face.

C. The frame face must be flush against the sides and top of the cupboard.

Step 5:

- Check that the following is true, and note the details:

    - The door back is ½-inch hardwood plywood. Note that hardwood plywood has a front and back.

    - Hardwood plywood has a face and back decorative covering made of hardwood instead of softwood like cedar or pine. It is stronger than solid wood and comes in varying species.

    - The door back is ½-inch hardwood plywood.

    - The back of the door is ½-inch hardwood plywood.

    - The front trim on the door is referred to as the "bender board. Note that the bender board is specifically good at fitting tight corners.

    - If you do not have a bender board, use ¼-inch craft boards for the trim.

Step 6:

- Saw the horizontal and vertical planks first, then cut to size and attach them with ⅝-inch brad nails.

Step 7:

A. Saw the diagonal boards to 1½ inches wide each. Mark them with a pencil to fit beforehand.

B. Attach them with ⅝-inch brad nails and glue.

Step 8:

A. The cupboard top is an edge-glued panel. Note that wood pieces glued along their edges to create products of greater width.

B. Adhere the top. After that, taper all the shelf supports. For the "tapering jig," you can just use a 2x4 clamped at a slight angle to the table saw sled. Drill a hole near the end to clamp it and cut the end of the 2x4 at the appropriate angle.

C. Using a push stick, hold down the wood and cut it.

D. Taper the shelf supports. Tapering jig is a 2 x 4 clamped at an angle to the sled of the table saw, with a drilled hole by the end to clamp. Then, cut the end of the 2 x 4 at the appropriate angle.

E. Use a push stick while you cut to hold down the wood.

F. Edge-glued panels usually come ready to varnish.

Step 9:

A. The cupboard back is ¼-inch hardwood plywood.

B. Cut the cupboard back to size and adhere it to the back using ¾-inch brad nails or staples.

Step 10:

A. The shelf planks will all be crafted with ¾-inch hardwood plywood. Make three.

B. Adhere the shelves using glue on both the mortise and tenon underneath each shelf plank. If you have shelf pin holes, make the planks a ¼ inch shorter, so there is room for the shelf pins on each side.

C. Craft the drawer supports. Using ¾-inch plywood and a table saw, cut to size.

D. Connect the top board mortise, then adhere them to the side tenons using wood glue.

E. Connect the mortise and tenons of the bottom and top of the vertical plank.

F. Do the same thing on the opposing side of the cupboard. Note that these are the planks your drawer slides will be attached to.

Step 11:

A. At last, craft your drawers.

B. Use a ¾-inch plywood for all sides of the drawers. Note that you can also use 1 x 4 planks.

C. Adhere each corner using 1¼-inch brad nails and wood glue.

D. Each drawer has a base plywood of ½ inch.

E. Using wood glue and 1¼ brad nails, attach it to the box.

Chapter Six: Project 4—Dining Table and Chairs

You have already made a dining table in the beginners' guide and chair in this book's first project. The unit of dining table and chairs is essential to understand scale and proportion. Making one object is easy, but making more in reference and correspondence to each other needs strong imagination.

Make a detailed sketch of your unit. If possible, you can use a computer-based software to create a 3D model that will help you view your ideal image and placement in a room. You can make it dynamic and contemporary in terms of design. You can also add a stool for seating arrangements for a kid.

If you are enjoying quality entertainment and dining with friends and family in your home, your dining room needs to have a comfortable atmosphere. Unfortunately, if the furniture is cramped or the chairs make people fidget during a meal, you will basically be unsatisfied. Sometimes, the dimensions of a dining room can be too limited, or the chairs are a hindrance. If so, it is time to put your intermediate skills of Japanese joinery to work. Maybe the shuffle to buy a new dining set seemed unappealing to your personal taste. Rather than acquiesce and buy something to "make do" or just have something new, you can set off on a new adventure in Japanese joinery. Follow the directions in this chapter to create a new and beautiful dining room table and chairs.

Many people think crafting your own furniture is solely for the wealthy; however, this quality furniture will not need replacing, and it will give you the chance to have a perfect fit. Your uniqueness will make a clear statement. You will know it was crafted with care and has long-term durability, giving you a high return on your investment. By commissioning your own dining room set, you will create an air of sentimentality for the whole family and all your friends, on top of it being a great conversational piece.

Selecting the Species of Wood

Mixing together different tones of wood is perfectly acceptable. However, as a starting point, pick a dominant tone to help you decide on other pieces to compliment the environment. For instance, if your floors are wooden, that would be the dominant tone. If you do not have a wooden floor, pick a dominant tone that matches the largest piece of furniture in the room. The creative goal here is continuity. Try to tell a story by noticing fine details, such as wood grain, undertones, and finish.

Match undertones: For example, many pieces of antique wood are in a warm mid-tone. Notice your dominant wood tone to see if it is warmer or cooler, and then try to continue that tone.

Playing with contrast: While it may seem contrary, mixing tones of wood is encouraged. Lightly colored floors may be complemented by a dark toned coffee table. Contrasting tones adds visual regard and provides more depth.

The finish: If you have mixed tones, you can show progression by using similar finishes. For instance, if you have a glossy floor or dining room table, follow suit with matching chairs that have a glossy finish.

Texture: Using different types of wood for furnishings does not simply mean to choose different wood stains or species of wood. It can also mean paying attention to the texture of the wood. Wood can be rough, smooth, knotty, weathered, or polished. Using wood with different textures can make for an interesting contrast in your decor. If your wooden floors are smooth, you may liven things up with a table you crafted with hand-scraped designs. Or you may compliment your smooth wooden floors with a knotty pine desk. As a tip, there can be such a thing as overkill when choosing your wood designs. It is best to stick to no more than a few different types of wood.

Accents: Breaking up your wood components with a nice rug can have a significant impact, especially if your wood tones are all the same. For example, if all of you have washed-out wood tones throughout the room, an area rug with deep colors will compliment your furniture.

Ahead of your project, place tape on your floor to see the size you would like for your table. Then, lay out the materials you are going to use to build the table, making sure you have the side you want to eat on face down (you are building it upside down).

Step 1 - The Shape you want

Decide if you want rectangular, oval, round, or square taking into account the shape of your dining room.

Step 2 - Material Selection

A. 4 X 8 Foot sheet of plywood ¾ inch plywood. Easy to cut and round of edges or corners. For a normal family size table.

B. *Material for legs*

- 2x4s
- Glue
- Clamps
- Drill & countersink bit
- Wood filler
- Table Saw
- Sander
- Sandpaper 120 grit

Wooden chairs are beautiful and timeless, as long as they are crafted to be comfortable. The chair needs to flex when sat upon. This chapter will step-by-step instruct the readers about a simple chair's function. The backrest is the most important feature of the chair. Most of all you will have a completely customized look for less than a fraction of the cost it would be to have one made. This chair will add a uniqueness to your home so you can avoid a generic looking style. You can take pride in knowing it is a one of a kind chair, not one a neighbor can come and see and then run out to buy. In this chapter the size, angle, proportions, and measurements will be covered along with discussions about the process, the angles, and various joints necessary to work the fittings. Finally, the importance of staining to give a finished look will be discussed. Crafting your own chairs puts the power of good quality and the look of costly finishes in quality furniture much less expensively than you would be at a store. By building your own furniture, you can add more support than anything you can find in a store. Also, your chairs will be made of 'real' wood which is made to stand the test of time. Choosing the right materials for your craft will determine its test of time. Knowing your own handiwork allows you to easily modify or fix something later.

Step 3 -Chairs

   Construct the legs

A. Start with cutting the 2x4s to length and apply plenty of glue to the front of one of the 2×4 pieces.

B. Place the other 2×4 on top and move it around to create a suction to keep the top piece from sliding. Clamp them together until dry.

C. Before heading to the router table, lay out the mortise locations on the workpieces, but mark on the faces opposite where the mortises will go. This way you can see the marks as you rout the mortised face.

- Make a simple box to how the wood bar vertically for when you route the tenon in the head.

- On the box top but an inlay board with a holed to guide the router and shape the tenon.

- Decide the width of your mortises and tenons.

- The tenon should be ½ the length of the short side of the wood bar section.

- Your router bit diameter should be of equal length.

- Use the same router bit to make the mortises and tenons.

- the mortise hole

- the tenon tongue.

- The tenon rail will fit into the hole you cut in the corresponding piece. Make sure to cut the tenon to fit your mortise hole exactly.

- The mortise has shoulders that seat the joints as it completely enters the hole of the mortise.

D. For square legs use the table saw with the blade as far raised as possible.

E. Run each leg through twice with the rip fence set at 3-14 inches from the blade.

F. Sand out all milling marks, scratches, transition marks from the manufacturing process.

G. Trim the other side of the leg assemblies with the rip fence on the saw set at 3 inches.

H. Attach them underneath the table using screws and wood glue.

    As a side note, if you would like to have a less formal dining room table instead, you can use large cable reels. They are the perfect height and strong enough to support the table. You can usually get them for free. Get two 29 inch reels and simply rest your table on top or coat with glue and attach your table top.

Design a few sketches before you start. Understand the measurement and purpose.

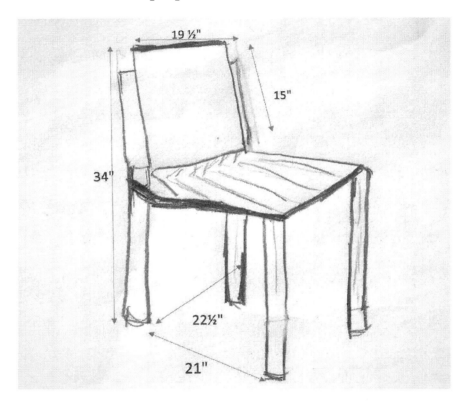

Chair dimensions

Materials:

1. 1x2" 8ft. 1 piece.

2. 2x2" 8ft. 2 pieces.

3. 2x4" 8ft.1 piece.

4. 1x10" 36in. 1 piece.

5. 3/8" dowel rod 1 piece.

6. Wood glue.

7. Wood filler.

8. 120 grit sandpaper.

9. Primer.

10. Wood conditioner.

11. Outdoor wood finish.

Wood Cuts:

1. 2 – 2×4 at 30"; Cut the back of the chair.

2. 2 – 2×2 @ 19 1/2"; Cut the supports for the chair back.

3. 2 – 2×2 @ 18"; Cut the front legs.

4. 2 – 1×3 @ 17 1/2"; Cut the side apron.

5. 1 – 1×3 @ 16 1/2"; Cut the front apron.

6. 1 – 1×3 @ 18"; Cut the back apron.

7. 5 – 1×4 @ 19"; Cut the seat board.

8. 5 – 1×4 @ 15"; Cut the back board.

Step 1:

Back legs (see diagram below): Mark your chair back legs on the 2 x 4. Be mindful of the back legs at 1½ inches wide in its entirety. One leg will take up the whole 2 x 4.

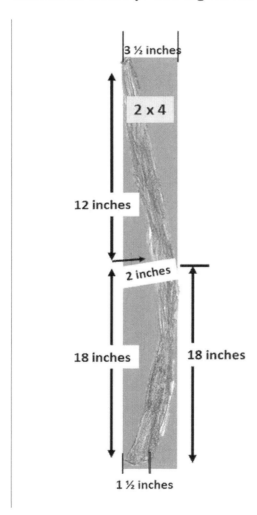

Step 2: Start with building the *back apron* and working your way to the front using a mortise and tenon technique.

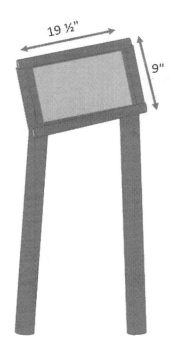

- 2 – 2×2 @ 19 1/2"; Cut the back supports from these planks.

- Connect the back support by constructing the mortises, and use tenons to connect the back support to the back legs.

Step 3: *Front legs and seat.* Start building the back apron and work your way to the front: *(See figure below)*

- 2 – 2×2 @ 18"; Cut the front legs

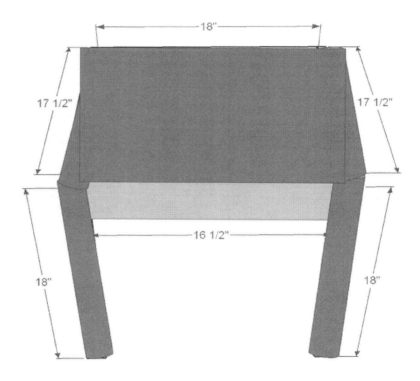

Step 4: Assemble the chair frame. Use a level on the side apron and connect the front of the chair to the back of the chair with mortise and tenons *(see figure below)*.

Step 5: Connect your mortise and tenons of the seat boards mindfully to the front and back aprons. Leave a ½-inch gap between each seat board.

Step 6: Adhere the back boards to the back supports, as you did with the seat boards.

Step 7: Use wood filler for any cracks or holes. Give your new chair a good sanding and finish or stain with your creativity. As a tip, before staining, be sure the chair is free of any dirt or anything sticky. Use a tack cloth in a mix of warm soapy dishwater to clean it off. Rinse with a damp sponge and let it dry all the way.

Step 8: Wipe the surface and legs of your table with mineral spirits, and then apply two coats of stain. Next, apply two coats of semi-gloss polyurethane. Let dry, and you're done!

The Importance of Burnishing

Well you have finished crafting your dining table and chairs, but is it really finished? If you don't want to leave it as raw wood, then it is not finished. There are a couple very important reasons why you burnishing your chair is a good idea.

First of all, wood is a permeable material and very proficient in soaking up just about anything that is spilled on or weathered into it. Finishing your project can seal the timber, keeping oils, chemicals, dirt, and dust out of the wood's grain. You want it to look fantastic for years to come.

The second important reason for burnishing your woodcraft is its grain. Although all timber has grain, some is more vibrant than other species of wood. Burnishing will liven up your wood and add color to the grain. However, it needs a clear-coat application after staining to protect the wood. There are a plethora of finishing types, and some can be used along with others or by itself. There isn't a right or wrong answer here because you will come to find your favorite after experimenting with different looks, desired effects, and application ease.

Paint is a basic finish, but not usually used for covering a beautiful piece of timber. However, there are some semi-transparent paints that, with a very thin coat, will help the grain to exhibit a dramatic effect. It is commonly used on plywood projects. Stain is applied over natural woods to accentuate the grain while still adding some color. Stain can come in dark brown, light brown, red, and other colors. It adds warmth and style to a piece, and it may look quite simple without it.

Clear Coat Finishes

Wax: Easily applied by rubbing and buffing to a shine. It is most effective on fine-grained woods that have already been sanded smoothly. It does not work well on rough-grained timber, such as ash or oak. Wax can also be used as a finish on top of other finishings to repel moisture.

Oils: Also easily applied to finished projects by loading it on and polishing to a dull sheen. The downside of oils is they do not make a shiny finish and can stay sticky for quite a while. However, if this is the way you want to go, you have a few options:

- *Linseed oil:* Can add a richness and warmth in color and increase its moisture resistance.

- *Tung oil:* Doesn't shine much, but does protect the wood.

- *Mineral oil:* In food grades, it is good for tending to children's toys and any kitchen appliances.

Shellac: A traditional finish, not commonly used anymore mainly because it doesn't do much in the way of protecting against moisture. It is a resin dissolved in alcohol and brushed onto the surface of the wood. It can be applied to add a natural, amber-looking warmth to your piece.

Varnish: Can be sprayed or smeared on. It does hold up to moisture and is much more durable than other types of stains, such as shellac. Lacquers dry much quicker than any of the oils, this finish dries hard and durable. You don't have to recoat with lacquer, but because it is so hard, it is subject to chipping and cracking with age.

Polyurethane: Gives a strong finish and is a liquid plastic. It comes in matte and glossy finishes and can be applied by brushing, wiping, or spraying. Be sure to use throwaway towels and even disposable brushes, as a solvent has to be used to clean any tools coated with polyurethane. It takes a while to dry, making it vulnerable to collecting dust, so keep this in mind when choosing where to finish your piece.

Polyacrylic: Water-based and dries fast to a durable finish. You can clean your tools with soapy water, so it is easier to use than polyurethane. It is also applied via cloth, brush, or spray.

Chapter Seven:
Project 5—Dynamic Bed with Two Dressers

Crafting a dynamic bed and accompanying dressers will allow you to use your Japanese joinery skills and be directly involved in the composition, materials used, and design process. That means no person on the planet will have a bed and set of dressers exactly like yours. A one-of-a-kind bed and dressers set will elevate your interior decor and give you a great sense of pride in its ownership. Controlling the design with your custom joinery has the benefit of you owning a handmade piece that fits your existing furniture. You will not have to compromise on the perfect style, since you can match other pieces you make as well. By making the set yourself, you need only add any adornments or special features you desire. For instance, if there is a joinery element you want to have on your bed, you need only to add them. If you want storage under your bed, you will have it custom-designed by you.

A dynamic bed means a bed with a bold and contemporary design. The shape could be square, octagonal, or pentagonal too. You can continue with the basic rectangle shape too. It is important to understand the complexity before you start, however. Seek the possibilities of making it interesting and not mundane one. The chests can be simple, which will flank the bed. Plan it in reference to the sleeper; it could be for kids, elders, or adults.

Make the design of both the bed and chest. First, finish the bed construction, and then follow it up with two chests to maintain the scale and proportion. Devise an accurate strategy and start the work with minimum effort. Remember to record the time whenever you take up a new project and read the entire plan prior to starting this project. Make sure to craft safely and smartly. Build on a level and clean surface, free of debris or bumps. Check for squares after every section. Use only straight boards and always pre-drill your holes before screwing. For a stronger hold, use finishing nails and glue, making sure to wipe excess glue so it doesn't stain. Most of all, have fun.

Materials:

- 2 x 6 (2; stud length).
- 2 x 4s slats (10; ten feet long).
- Five 2 x 4s (5; stud length for headboard and remaining slat piece).
- 1 x 3 (1; 8 feet long).
- 2-inch screws.
- 2½-inch pocket hole screws (2).
- Wood glue.

Wood Cuts

- Headboard
  - 2 x 6 at 79 inches (6).
  - 2 x 4 at 79 inches (1).

- Back Cleats
    - 1 x 3 at 31½ inches (3)
- End
    - 2 x 6 at 61 inches (1).
    - 2 x 4 at 65 inches (1).
- Sides
    - 2 x 4 at 79 inches (2)
    - 2 x 6 at 79 inches (2)
- Slat system
    - 2 x 6 at 79 inches (3).
    - 2 x 4 at 58 inches (11; 4-inch gaps between slats).

Step one: The headboard needs to be attached to the wall so it does not bowl forward. Use a cleat on the back of the headboard *(see diagram below)*.

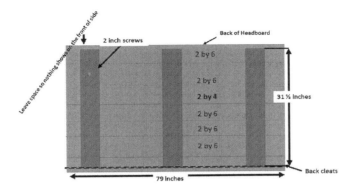

Step 2: Construct two side rails by adhering the 2 x 6 upright pieces to the 2 x 4 platform pieces using one 1½-inch pocket hole and 2½-inch pocket hole screws. Drill pocket holes on the ends at 1½ inches, so you can attach the headboard and footboard later.

Step 3: For the footboard, build the platform top piece so it over extends the upright piece by 2 inches on the ends, keeping everything flush on the back *(see illustration below)*.

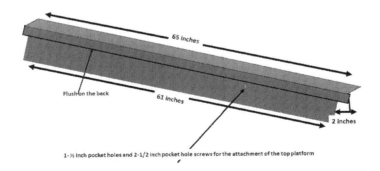

Step 4: Stain the sides and end pieces after they are constructed, and then join the side pieces to the end with 2-½ inch pocket hole screws.

Step 5: Place studs in the wall to keep the headboard in place and prevent it from warping. Adhere the headboard through the pre-drilled pocket holes with 2-½ pocket hole screws.

Step 6: Put the slats in the frame and screw them down.

Step 7: To finish, use wood filler for all holes and leave to dry. Add more if needed. After the filler is completely dry, sand the wood in the direction of the grain using 120 grit sandpaper. Vacuum the bed after sanding and remove any sanding residue. Wipe bed with clean dampened cloth.

Step 8: Apply a sample coat of your chosen finish on a scrap piece to make sure it sticks and and the color is even. Use a wood conditioner or primer if desired. Piece together and ensure color evenness and adhesion. Use a primer or wood conditioner as needed.

Bedroom Chests

Materials:

- ¾-inch plywood 48 x 96 inches (1). Your choice of hardwood: 2 pieces for glide supports, drawers, sides, top, and bottom.

- Hardwood: ¾-inch thick, 12 x 96 inches for the fronts of the drawers (1).

- Hardwood: ¾-inch thick, 3 x 96 inches for the top of the frame (1).
- Hardwood: ¾-inch thick, 2 x 96 inches for the frame face (3).
- Hardwood: ¾-inch thick, 5 x 96 inches for the dividers in the drawers (1).
- Plywood: ¼-inch thick, 48 x 96 inches for the bottom of the drawers and panel on the back (2).
- ¼ lb of 1¼-inch wood screws.
- ⅛ lb of ½-inch wood screws.
- ¼ lb of 1¼-inch Kreg screws.
- Sandpaper.
- Glue.
- Wood filler.
- Polyurethane finish.

Tools

- Power drill with a tapered bit.
- Table saw.
- Band saw.
- Sandpaper at 120 grit and a 320 grit block for finish.
- Tape measure.

- Router with round-over bit.

- Pencil.

- Corner/arc template.

- Miter saw.

- Pocket hole joinery system.

- Screw gun.

- Bar clamps.

- Gripping clamps.

Step 1: Cut the Dresser Box Pieces

Using a ¾-inch plywood stock, cut one piece in half, so you have 24 inches. Then, crosscut the 24 x 96 inch in half, and you will have the two sides at 48 inches. With the other half (24 x 96 inches), cut out the bottom panel at 24 x 34½ inches *(see illustration below)*.

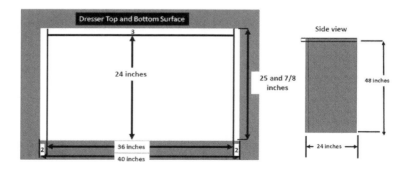

A. *Top panel*: From the plywood stock, cut one piece, 24 x 36 inches.

B. *Pick the top, side, and bottom panels*: On the back edge, make a slot cut (slot cut is also referred to as a daddo cut) a ¼ inch deep by ½ inches to gain the ¼-inch back panel.

C. Pick the 3-inch hardwood stock: Cut one 42 inch piece for the *top frame front panel*; make 45-degree cuts on the ends.

D. Pick the 2 inch hardwood stock and cut two 25⅞ inches for the *top panel sides*; cut one end on each at 45 degrees.

E. Pick the *side panels, top, and bottom* and drill pocket holes down the edges of the panels. Make sure you drill the hole to adhere to the front frame.

F. Pick the ¾ inch plywood stock and cut 8 pieces at 2 ⅞ by 23 ½ for the *slide supports*.

Step 2: Craft the Dresser Box

A. Make the *dresser box* with the four pieces that are used for the *top panel*, found in step one. With glue, pocket hole jig them together. Next, use a band saw and round the front two corners. Then, use a router with a roundover bit to round off the front and side bottom edges (*not* the back edge). Sand them down.

B. Pick the *eight slide supports* and *two side panels*. Use 1¼-inch screws and glue to adhere the supports on the two side panels. Countersink the screws using the dimensions labeled in step one.

C. Pick the *bottom panels, top panels, and two side panels*. Use glue and pocket holes together. Take not of the 2-inch placement of the bottom panel.

Step 3: Craft and Adhere the *Front Frame* Using the 2-Inch *Hardwood* Material

A. Cut two 1½ x 48 inches from the 2-inch material.

B. Cut five 2 x 33 inches from the 2-inch material.

C. Drill pocket holes in the select cut pieces, then sand them down.

D. Using your pocket hole screws and glue, join the pieces.

E. Picking the *front frame* and using your pocket hole screws and glue, join the frame and cabinet.

Step 4: Craft the Drawers Using the Plywood and Hardwood Stock

A. Cut the *front and back panels* from the ¾-inch plywood for the three bigger drawers. Cut six pieces at 9½ x 30½ inches. Next, cut a ¼-inch slot by ⅜ of an inch deep to get the ¼-inch *bottom panel*.

B. Cut the *side panels* from the ¾-inch plywood. For the **three bigger drawers**, cut six pieces at 9½ x 23 ⅞ inches. Next, cut a ¼-inch slot, ⅜ of an inch deep to get the *bottom panel*.

C. Cut the front and back panels from ¾ of an inch plywood for the smaller top drawer. Cut two pieces at

7 x 30½ inches, then cut a ¼-inch slot, ⅜ deep to get the ¼-inch bottom panel.

D. Cut the side panels from the ¾-inch plywood for the smaller top drawer. Cut two pieces at 7 x 23⅞ inches. Cut a ¼-inch slot, ⅜ inches deep to get the ¼-inch bottom panel.

E. Drill two holes in each front and back panel on counter sink holes (center) to get the *drawer center divider*.

F. With the side panels, drill three counter sink holes on each edge.

G. Cut four bottom panels from the ¼-inch plywood stock at 31 x 23⅛ inches. Do so without gluing or fixing. Test the drawers on one panel to make sure your measurements are correct prior to butting the other three bottom panels.

H. Cut four center dividers at 4½ inches by 22⅜ inches from the ¾-inch hardwood stock.

- Sand all the pieces.

- With glue and 1¼-inch screws, craft all the drawers from all of the pieces.

Step 5: Drawer Slides

A. Attach the slides to the drawers and chest glide support braces.

Step 6: Craft and Adhere the Front of the Drawers

A. Cut three larger drawer fronts from the 12-inch hardwood stock at 11 x 34 inches.

B. Cut one small drawer face at 8½ x 34 inches.

C. Use a router on all edges—even where the finger grabs are—and craft the drawer faces.

D. Center the faces and clamp them on the drawer faces; pre-drill and adhere with 1¼ inch screws.

Step 7: Varnish the Dresser!

A. Cut the back panel from the ¼-inch plywood at 35¼ x 46⅞ inches using the ½-inch screws. Fill any cracks or holes with filler and give all surfaces a final sanding.

B. Put at least three coats of polyurethane on the dresser.

Chapter Eight: Maintenance of Wood and Japanese Lacquering

Japanese lacquer is specific when it comes to maintenance due to the nature of its construction. It is necessary to have a good understanding of how it is made and what elements may contribute to its degradation over time.

Natural lacquer has been produced in Japan for thousands of years and was used as a protective coating for various household times, musical instruments, weapons, and various other cherished items. When it is raw, in Japanese, *urushi* is a milky tree sap, which is then drawn, heated, filtered, and stored. The lacquer's application is a thin layer to a prepared element, such as wood and leather. It is then allowed to preserve in a warm case, free of dust, for more than a day. After polishing, a second layer is applied. Depending on the artist, the piece of work may only have a couple layers or even a hundred layers (Lauern, 2013).

Kijiro Lacquering

Kijiro lacquering is a transparent coating that exposes the grain of the wood for its beauty. As a base, it has a clear to amber coating, with its charm deeply expressed. If you add more coats, the depth will continue to advance, but the grain will be less visible.

Kamakura-Bori Lacquer

Kamakura-Bori lacquer has been a tradition in Japanese joinery dating back to the eleven hundreds. It was used to lacquer altars and Buddhist statues. Finished pieces are called *kamakura-bori*. This type of finish has intentional chisel markings left to liven patterned sections. Another special technique is to scatter black powder on a reddish lacquered surface, so the patterns are noticeable against the darkened background.

Here are the steps to follow:

1. Use knives (of all different shapes) to chisel marking on the surface of your project to accentuate the patterns on the surface.

2. Use the lacquer also as an undercoat. Apply one coat to the whole surface, including any engraved lines or chisel marks.

3. Then, scatter some carbon dust, polishing carbon dust, and polishing talc to maximize the lacquer's effect on the carved surface. This process accentuates the carved, uneven surface.

4. For the middle coat, apply two layers of black *urushi* lacquer, making sure it does not pool into any of the dents made by your chisel. When it's dry, sand the surface.

5. For the final coat, use a clear Japanese lacquer, and before it is dry, sprinkle *makomo* powder on the surface.

6. To finish, apply the lacquer again over the entire surface. Let dry and polish to complete the *kamakura-bori* project!

Fuki-Urushi

Fuki-urushi is a lacquer where the natural glow emphasizes the beauty of the timber's grain.

1. Make sure all fingerprints and dust are removed from the surface of the wood.

2. Apply the lacquer with a brush.

3. Allow the wood to soak up the lacquer, as it will make the grain pattern stand out.

4. Rub it into the wood and let it dry.

5. Carefully polish with sandpaper.

6. Repeat step 2 five times. Note that this process may take a month or more to complete.

Wajima Kirimoto's Sensuji

Wajima Kirimoto's Sensuji technique is created by drenching a cloth with the lacquer and applying it to parts of the wood piece that can be damaged easily. A special brush that creates a bold pattern with many streaks done with jinoko makes for a tough surface and gives the wood an antique feel.

The Honkataji Technique

The *Honkataji* technique is where lacquered clothes are applied to the wood in its natural state and on the areas that are easily damaged, such as the tops of furniture. Then, *jinoko* powder is used as the base coat, it is polished, and a middle coat is added. Once all that is done, a top coat is added. The finished piece will show nine layers of lacquer, leaving the piece always able to be repaired, if damaged.

Caring for Japanese Lacquer

Producing lacquered wood pieces is in depth and takes time and patience. Many special tools and steps are involved in the process. Lacquer can also be colored with minerals or manufactured pigments. For example, iron oxide and cinnabar are combined to make red lacquers. It may also be used in the natural clear state. It can be decorated by adding mother of pearl or metal and polished until the decoration is revealed. Grades of powdered silver and gold with great care are set into the wet lacquer for detail or background effects. However, even though lacquer is hard once manipulated, it can still be damaged. When a lacquered object is in an environment with fluctuating degrees of humidity, lifting and cracking can occur. Therefore, these objects should be handled carefully and with mindfulness when placing them near high moisture areas, cooling and heating vents, or drafty areas. Close attention should be given to how a lacquered object is handled or filled with heavy objects. Too much light can advance a lacquer's degradation, making it lose its durability. Thus, you should make sure you do not expose it to direct sunlight or UV lighting. Watch out for window light by using curtains or filters.

Fingerprints, scratches, and other smudges can damage the lacquer surface. Wear cotton gloves and dust frequently, as some dust particles are abrasive. Never submerge lacquered items in water or use commercial dust removers, such as Endust or Pledge, which can ruin the finish. Lastly, it is imperative to know that Japanese lacquer is not the same at Western shellac. Japanese lacquer is frequently imitated, so be aware; if you are unsure of the lacquer's origin, note that it may be shellac. If it is Japanese lacquer, remember that it is fragile.

To restore the lacquer's appearance, make an overall assessment first, looking for any issues, then use carnauba wax to polish the piece. Apply the wax with one soft cloth, and then polish the item lightly with another clean soft cloth. Do *not* wax pieces that are in good condition. Wood pieces which are lacquered can become insect-infested. A sign of insect infestation includes pin holes surrounded by frass (a sandy-like substance). To remedy this, freezing or fumigating should work.

The best way to care for Japanese lacquerware is to use it periodically because it extends and maintains the integrity of both. Japanese lacquer contains moisture so it shines more after applied. This does not happen with chemical products and therefore, it is a testimony to its authenticity. Be careful handling metals or ceramic products so as to not scratch the fine lacquer.

Dos and Don'ts for Japanese Lacquer

- *Don't* microwave.

- *Don't* place it in a dishwasher.

- *Don't* leave it in the refrigerator for more than an hour or two.

- *Do* wash by hand with warm, soapy water.

- *Don't* use abrasive sponges.

- *Don't* stack lacquerware.

- *Don't* use metal silverware on lacquerware.

- *Do* handle with care.

Most individuals can easily feel that a certain piece of furniture has sentimental value, whether it is the dining room set you learned how to craft here or the bedroom set. When you pass these pieces down through the generations, they will bring with them the memories and stories of your past, making them even more valuable to your family. Hardwood furniture made with the Japanese joinery is crafted to be heirloom quality, for it can be revered for future generations. Handcrafted hardwood furniture has a great advantage in its appeal and endurance. The tones and grain patterns of real hardwood bring character and warmth to any room.

At the intermediate level, your skill in finishing will never be matched with veneer. Even when it ages, it becomes part of a very "in-style," rustic look. An old dresser, for instance, can add a stately charm to the master bedroom. As a crafter skilled in Japanese joinery, you can look at a log that has been cut in half and see the graceful diagram of a dining room table beneath the wood's bark. When you see a piece of high-quality wooden furniture with deep dents and other shown damage, you will immediately think of its character, how to bring it back to life, and be eager to start the process. Besides having the ability to become an heirloom, there are many reasons handcrafted furniture is a smarter buy than mass-produced items.

Durability: Furniture built with Japanese joinery is made to tolerate years and years of wear and tear while keeping its natural beauty. It can be finished with protective sealants, making it scratch, stain, and moisture-resistant. Most manufactured furniture is made from plywood or even particle board and quite simply cannot match the longevity of handcrafted hardwood furnishings.

Renewability: Trees of hardwood are renewable resources and add to a healthy environment while they are growing. When harvested, trees drop seed to replace themselves. Unlike mass-produced furniture sets that contain preservatives like formaldehyde or fillers, the hardwood used in crafting the pieces set forth in this book are friendly to the environment. The crafter decides which stains and varnishes they will use and where they are sourced. Pieces can be sourced yourself or from local lumber shops, reducing the cost of transportation and supporting the local economy.

Beauty: Every species of timber has characteristics and grain pattern unique to itself. Even within the same species, wood grain patterns vary from section to section, making every piece of furniture you craft one of a kind. Finishes and stains further add to the luster of the wood.

Customization: Unlike manufactured and mass-produced furniture, hardwood, handcrafted pieces can be created to your own specifications and needs. With options for different stains and finishes and the skill of Japanese joinery techniques, your furniture can be customized to fit your home, style, and preferences.

Restorable. Although hardwood furniture is built to last for generations, the ability to restore it down the road allows it to take on a new life, fit changing style preferences, or match the decor of a new household. Manufactured furniture made of composite material cannot typically be restored and is often discarded when it falls apart or shows significant signs of wear.

Retains its value: Although crafting hardwood furnishing with Japanese joinery techniques can typically be more expensive up front, it certainly holds its value. It retains much more value than manufactured furniture, and down the road, it will end up saving you more money than buying manufactured furniture.

Hopefully by studying this book, you will have increased your skill level and expertise in Japanese joinery. Your knowledge and understanding of Japanese joinery woodworking and how to construct your work has grown exponentially, and hopefully, this book has given you more confidence.

By completing this second book in the series of the Japanese joinery, your skill level has grown, along with the complexity of the pieces you choose to craft. Saving money is not the main impetus for finishing your projects, but it does deserve to be noted because of the expense that you can find with solid wood furniture. Most of all, you will appreciate your work and creativity much more than bringing something home with your debit card. Being involved in designing and creating your own projects with Japanese joinery will lend to it a deeper meaning, as you will know how much work and detail went into every aspect of your creation, leaving you with a great sense of pride and satisfaction. In the future, you will be able to use your Japanese joinery skills to build anything you want that will meet your specific needs.

Anyone who has shopped for furniture has felt the agony of trying to find the right piece to match your vision or accommodate your intentions. Always keep in mind that caring for your tools and keeping them in working order—such as sharpening your knives and wiping them with oil—will be the most important part in your journey with Japanese joinery. Vegetable oil becomes tacky, so you will want to use an oil, such as walnut. Your joinery knowledge should now make you more confident in planning and fitting your furniture into any space perfectly with the exact dimensions necessary, in the exact color, and with any bells and whistles your creative mind desires.

References

Brownell, B. "A History of Wood and Craft in Japanese Design" The Journal of the American Institute of Architects, 2016. https://www.architectmagazine.com/technology/the-history-of-wood-and-craft-in-japanese-design_o

Butler, L. "Patronage and the Building Arts in Tokugawa Japan", Early Modern Japan. Fall-Winter 2004 https://kb.osu.edu/bitstream/handle/1811/5820/V12N2Butler.pdf;jsessionid=6104A29BFC3D655AA6BCD1605C6DDE80?sequence=1

Hover, B. "How To Find Pearls In Burls". Wood Magazine.com, 2020. https://www.woodmagazine.com/materials-guide/lumber/wood-figure/how-to-find-earls-in-burls

Hyogo-ken Chijitoroku Ryokogyo. "Japanese Carpentry". Japanese Guest Houses 2020.

https://www.japaneseguesthouses.com/japanese-carpentry/

Kiyosi S. "The Art of Japanese Joinery". Tankosha, 1978. https://dokumen.pub/the-art-of-japanese-joinery-1nbsped-0834815168-9780834815162.html

Sennet, R. "The Craftsman", Allen Lane, London, 2008

Toshio Odate. "Japanese Woodworking Tools: Their Tradition, Spirit and Use" (Linden Publishing, Fresno, CA, 2006), p39

Hiroyuki Sugawara. "Oen" 2020. https://the189.com/feature/japanese-woodworker-hiroyuki-sugawara/

Keller, Seth. "Tools we Love: Japanese Ryoba and Mini Dozuki Handsaws". Woodworkers Guild of America, 2020.

Made in the USA
Las Vegas, NV
20 November 2024